Excel

YEAR 9

Problem Solving Workbook

ESSENTIAL skills

Get the Results You Want!

Allyn Jones

Reprinted 2017, 2022, 2023

Reviewed in 2025 for the NSW Curriculum and Australian Curriculum Version 9.0 changes

ISBN 978 1 74125 570 6

Pascal Press
PO Box 250
Glebe NSW 2037
www.pascalpress.com.au

Publisher: Vivienne Joannou
Project editor: Rosemary Peers
Edited by Rosemary Peers
Proofread by Barbara Bessant
Answers checked by Peter Little
Cover, page design and typesetting by DiZign Pty Ltd
Printed by Vivar Printing/Green Giant Press

Introduction

This book has been specifically written for the **Australian Curriculum**.

The Australian Curriculum (Mathematics) covers six areas:

- Number
- Algebra
- Measurement
- Space
- Statistics
- Probability.

Students are encouraged to develop proficiency with mathematical concepts, skills, procedures and processes, and to use them to demonstrate mastery in mathematics as they solve problems.

This book provides strategies for students when applying mathematical concepts in a variety of routine and non-routine problems of increasing difficulty.

Please go to page 9 for Contents

Please go to page 9 for Contents

HOW TO USE

KEY SKILL

44 Trigonometry: Bearings A

HINTS

- A bearing to a point is an angle, expressed in 3 digits, measured in a clockwise direction from north.
  e.g. What is the bearing of *A* from *B* and the bearing of *B* from *A*?

N
B 40°
N
40°
A

As 90 + 40 = 130, the bearing of *A* from *B* is 130°.
As 270 + 40 = 310, the bearing of *B* from *A* is 310°.

Reminder!

- Read the question carefully to identify what needs to be found.
- Re-read the question when you've finished your answer to make sure that the question has been answered and your solution makes sense.

Examples

A ship leaves Port Hedland and travels 380 km on a bearing of 335°. How far north of Port Hedland is the ship, to the nearest kilometre?

Solution

335 − 270 = 65

Let the distance be x.

$$\frac{x}{380} = \sin 65°$$
$$x = 380 \times \sin 65°$$
$$= 344.396\,9591\ldots$$
$$= 344 \text{ (nearest whole)}$$

S N x 380 65° P

∴ the ship is 344 km north of Port Hedland.

FOCUS on...

1. The question: Asks you to find the distance the ship is north of Port Hedland.
2. The information: Gives you the distance and bearing the boat has travelled.
3. Your working: Draw a triangle to represent the information. As the bearing is 335°, you need to subtract 270° to find the angle inside the triangle. Let the distance north of Port Hedland = x.
   From the given angle, the sides are the opposite and hypotenuse. This means you will use sin:
   $$\frac{x}{380} = \sin 65°$$
   $$x = 380 \times \sin 65°$$
   $$= 344.396\,9591\ldots$$
   $$= 344 \text{ (nearest whole)}$$
4. Your answer: Make sure you write the correct units: The ship is 344 km north of Port Hedland.

A plane takes off from an airstrip and flies 160 km east and then turns and flies 130 km south. What is the bearing of the plane from the airstrip? Give your answer to the nearest degree.

Solution

Let the angle be θ.

$$\tan\theta = \frac{130}{160}$$
$$\theta = 39.093\,858\,89\ldots$$
$$= 39 \text{ (nearest whole)}$$

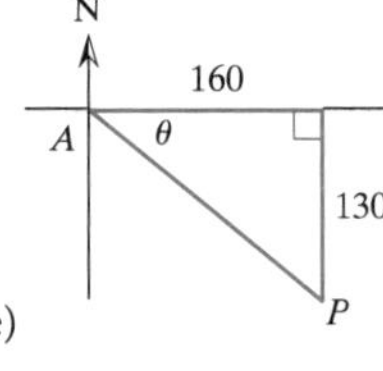

90 + 39 = 129

∴ the bearing is 129°.

FOCUS on...

1. The question: Asks you to find the bearing of the plane from the airstrip.
2. The information: Gives you two distances the plane has flown.
3. Your working: Draw a triangle to represent the information. Let the unknown angle = θ.
   From the unknown angle, the given sides are the opposite and adjacent. This means you will use tan:
   $$\tan\theta = \frac{130}{160}$$
   $$\theta = 39.093\,858\,89\ldots$$
   $$= 39 \text{ (nearest whole)}$$
   The plane has flown 39° south of east. This means you need to add: 90 + 39 = 129.
4. Your answer: Make sure you write the correct units: The plane is flying on a bearing of 129°.

108 *Excel* Year 9 Problem Solving Workbook

THIS BOOK

Key skill

- Each unit focuses on one key skill from each of four sub-strands of the curriculum, e.g. Trigonometry: Bearings A in the Measurement and Geometry strand.

Hints

- Students are given hints to help them solve the problems in the unit.
- These hints may also remind students of rules covered earlier in the book.

Reminder

- Students are given a reminder to always follow certain steps when they begin and end each question.

Examples

- Two examples are provided for each unit with worked solutions that students can follow.

Focus on ...

- A step-by-step method is provided for each example.
- These step-by-step methods are then used to solve the questions in the **Now try these!** section.
- Explanations and tips are provided in the methods to help students in their understanding and to encourage them to develop logical processes when solving problems in this unit.

A step-by-step guide to problem solving

Step 1: Focus on the question

- Read the question at least twice to make sure that you understand what you are being asked to solve.

Step 2: Focus on the information

- Concentrate on the relevant information. In some questions you may be given unnecessary information which will not be useful.
- Decide what you are looking for.

Step 3: Focus on your working

- There are many problem-solving strategies that you can use, including:
  - looking for a simpler problem of the same type
  - using easier numbers
  - working systematically
  - using the guess-and-check method
  - eliminating possibilities
  - drawing a picture or a diagram
  - drawing up a table
  - looking for a pattern
  - working backwards.

Step 4: Focus on your answer

- Make sure you write the correct units (e.g. minutes, metres) and ask yourself these questions:
  - Does the answer make sense? Is it a realistic answer?
  - Have I answered the question I was asked?

Now try these!

- All the questions in this section have been carefully written so that the exact same steps in solving them replicate the steps in the examples.
- The CHALLENGE question is a more challenging question where students will need to use slightly different steps to solve it.

Now try these!

1. A wall with a height of 3.2 metres casts a shadow of 4.8 metres. At the same time a tree casts a shadow of 16 metres. What is the height of the tree, to 2 decimal places?

2. 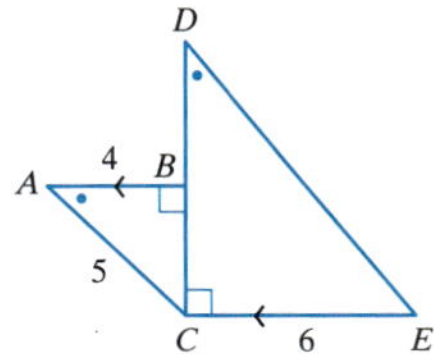

Find the length of DB.

3. A building of height 38 metres casts a shadow of 64 metres. A flagpole on the top of the building casts a shadow of 4 metres. How high is the top of the flagpole from the ground?

4. 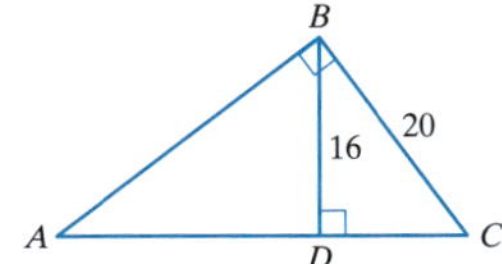

Find the length of AC.

5. Mia cuts a 4-mm diameter circular hole out of a piece of cardboard and holds the cardboard 45 cm from her eyes so that the moon just fits inside the circle. If the diameter of the moon is 3475 km, how far is the moon from Mia?

6. CHALLENGE

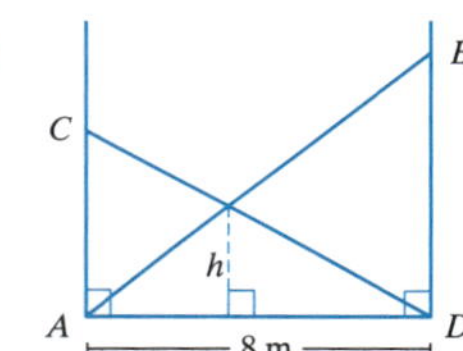

The diagram shows two ladders AB and DC, placed in a narrow alley. $AB = 17$ m and $DC = 10$ m and the alley is 8 metres wide. Find the height h, where the two ladders cross.

Answers pages 151–152

Revision Tests

- Pairs of revision tests appear regularly throughout the book to test students' understanding.
- Each pair has a test at both an Average and Challenging level of difficulty.

REVISION TEST 9 Level of difficulty—Average

1

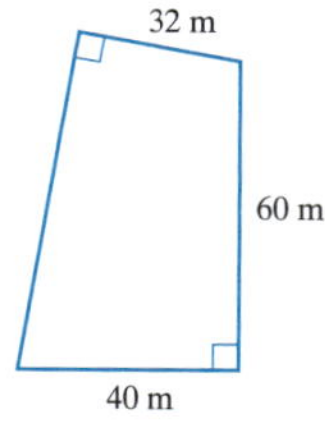

The diagram shows a paddock. If fencing costs \$8.25/metre, find the cost of fencing the paddock, to the nearest dollar.

2 Two jets leave from the same airport. One jet leaves at 3 pm and travels north at an average speed of 1000 km/h. The other jet leaves the airport at 4:15 pm and travels west at an average speed of 1200 km/h. How far apart are the jets at 5:00 pm, to the nearest 100 kilometres?

3 A rectangular prism has a surface area of 416 cm^2. The prism has a length of 10 cm and a height of 6 cm. What is the width of the prism?

4

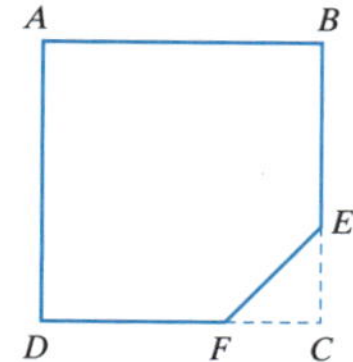

The perimeter of the square $ABCD$ is 80 cm. If $EC = FC$ and $EF = 4$ cm, what is the area of pentagon $ABEFD$?

5

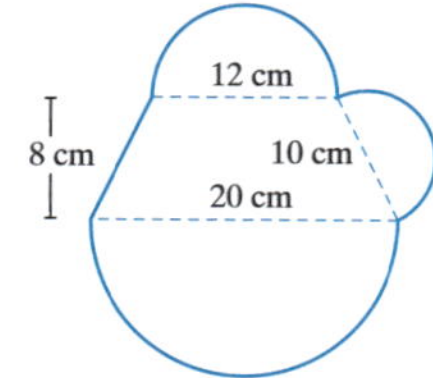

A shape is comprised of a trapezium and three semicircles. Find the exact area of the shape.

6 The curved surface area of a cylinder with height 18 cm is 540π cm^2. What is the total surface area of the cylinder, in terms of π?

Answers page 148

86 *Excel* Year 9 Problem Solving Workbook

Quick answers

- Students can quickly mark their work by referring to the quick answers in **bold** type.

Worked solutions

- Each question has a worked solution so that students can check incorrect answers or alternative ways to achieve the answer.

Worked solutions

Key Skill 1 (pages 10–11)

1. **$1331.20**
Total pay $= 25.6 \times 28 + 25.6 \times 1.5 \times 8 + 25.6 \times 2 \times 6 = 1331.2$
∴ her total pay was $1331.20.

2. **$21**
Let her hourly rate be x.
$756 = 18 \times x + 4 \times 1.5 \times x + 6 \times 2 \times x$
$= 18x + 6x + 12x$
$= 36x$
$x = \frac{756}{36}$
$= 21$
∴ her hourly rate was $21.

3. **$4575**
Total pay
$= 75 \times 42 + 75 \times 1.5 \times 6 + 75 \times 2 \times 5$
$= 4575$
∴ his total pay was $4575.

4. **4**
Let her double-time hours be x.
$813.2 = 21.4 \times 18 + 21.4 \times 1.5 \times 8 + 21.4 \times 2 \times x$
$= 642 + 42.8x$
$42.8x = 813.2 - 642$
$42.8x = 171.2$
$x = \frac{171.2}{42.8}$
$= 4$
∴ Tara worked 4 hours at double time.

5. **Jordan by $39.60**
Damian's pay
$= 13.2 \times 10 + 13.2 \times 1.5 \times 6$
$= 250.8$
Jordan's pay
$= 13.2 \times 8 + 13.2 \times 2 \times 7$
$= 290.4$
Difference $= 290.4 - 250.8$
$= 39.6$
∴ Jordan is paid $39.60 more than his brother.

6. **$289.50**
Let his hourly rate be x.
$492.15 = 12 \times x + 9 \times 1.5 \times x$
$= 12x + 13.5x$
$= 25.5x$
$x = \frac{492.15}{25.5}$
$= 19.3$
∴ his hourly rate is $19.30.
Next pay $= 19.3 \times 9 + 19.3 \times 2 \times 3$
$= 289.5$
∴ Leonard will earn $289.50.

Key Skill 2 (pages 12–13)

1. **$6019.25**
Taxable income $= 46\,920 - 2390$
$= 44\,530$
Tax payable $= 3572 + 0.325 \times 7530$
$= 6019.25$
∴ Amelia will pay $6019.25.

2. **$1549.80**
Total fortnightly tax paid
$= 788 \times 26$
$= 20\,488$
Taxable income $= 86\,300 - 2540$
$= 83\,760$
Tax payable $= 17\,547 + 0.37 \times 3760$
$= 18\,938.2$
Refund $= 20\,488 - 18\,938.2$
$= 1549.8$
∴ Cooper will receive a refund of $1549.80.

3. **$14 695.45**
Total deductions $= 670 + 504$
$= 1174$
Taxable income $= 72\,400 - 1174$
$= 71\,226$
Tax payable
$= 3572 + 0.325 \times 34\,226$
$= 14\,695.45$
∴ Mia will pay $14 695.45.

4. **$11.50 more to ATO**
Total income $= 38\,600 + 16\,460$
$= 55\,060$
Tax payable
$= 3572 + 0.325 \times 18\,060$
$= 9441.5$
Difference $= 9441.5 - 9430$
$= 11.5$
∴ Heidi still needs to pay $11.50 more.

5. **23.5% and 13.6%**
Jayden's tax payable
$= 17\,547 + 0.37 \times 9400$
$= 21\,025$
Jayden's tax % $= \frac{21\,025}{89\,400} \times 100\%$
$= 23.5$ (1 dec. pl.)
Bella's taxable income $= 89\,400 \div 2$
$= 44\,700$
Bella's tax payable
$= 3572 + 0.325 \times 7700$
$= 6074.5$
Bella's tax % $= \frac{6074.5}{44\,700} \times 100\%$
$= 13.6$ (1 dec. pl.)
∴ Jayden pays 23.5% and Bella pays 13.6% of their income in tax.

6. **$50.40**
Annual income $= 3200 \times 26$
$= 83\,200$
Tax payable $= 17\,547 + 0.37 \times 3200$
$= 18\,731$
Increased income $= 83\,200 \times 1.025$
$= 85\,280$
New tax payable
$= 17\,547 + 0.37 \times 5280$
$= 19\,500.6$
Extra tax $= 19\,500.60 - 18\,731$
$= 769.6$
Extra fortnightly tax $= 29.6$
Extra weekly income
$= 3200 \times 0.025$
$= 80$
Extra in pay packet $= 80 - 29.6$
$= 50.4$
∴ Ashley will receive an extra $50.40.

Key Skill 3 (pages 14–15)

1. **$94 600**
Xavier's income was between $80 000 and $180 000.
Let his income be x.
$22\,949 = 17\,547 + 0.37 \times (x - 80\,000)$
$5402 = 0.37(x - 80\,000)$
$x - 80\,000 = \frac{5402}{0.37}$
$= 14\,600$
$x = 94\,600$
∴ Xavier had a taxable income of $94 600.

2. **$78 450**
Total fortnightly tax paid
$= 670 \times 26$
$= 17\,420$
Tax payable from table
$= 17\,420 - 896.75$
$= 16\,523.25$
His income was between $37 000 and $80 000.
Let his income be x.
$16\,523.25 = 3572 + 0.325 \times (x - 37\,000)$
$12\,951.25 = 0.325(x - 37\,000)$
$x - 37\,000 = \frac{12\,951.25}{0.325}$
$= 39\,850$

126 *Excel* Year 9 Problem Solving Workbook

Contents

Number and Algebra

Measurement and Geometry

Statistics and Probability

KEY SKILL 1

Financial mathematics: Overtime penalty rates

HINTS

- A wage earner is paid on the basis of an hourly rate, e.g. $16.20 per hour.
- When overtime is worked the employee can receive penalty rates, such as time-and-a-half, double time, etc., e.g. Evaluate
  $19.53 \times 35 + 19.53 \times 1.5 \times 8$
  $= 917.91$

Reminder!

- Read the question carefully to identify what needs to be found.
- Re-read the question when you've finished your answer to make sure that the question has been answered and your solution makes sense.

Examples

Mia is paid an hourly rate of $15.20. In one week she works 28 normal hours, 6 hours at time-and-a-half and 4 hours at double time. What is her total pay for the week?

Solution

$$\begin{aligned}\text{Total pay} &= 15.2 \times 28 + 15.2 \times 1.5 \times 6 \\ &\quad + 15.2 \times 2 \times 4 \\ &= 684\end{aligned}$$

∴ Mia is paid $684.

FOCUS on ...

1. The question: Asks you to find Mia's total pay.
2. The information: Gives the usual hourly rate of pay and the number of hours worked at normal time, time-and-a-half and double time.
3. Your working: You need to add the three amounts she is paid. For time-and-a-half hours you multiply by 1.5 and for double-time hours you multiply by 2.
   $$\begin{aligned}\text{Total pay} &= 15.2 \times 28 + 15.2 \times 1.5 \times 6 \\ &\quad + 15.2 \times 2 \times 4 \\ &= 684\end{aligned}$$
4. Your answer: Make sure you write the correct units: Mia is paid $684.

Jack's hourly pay is $17.30. In one week he receives $986.10 when he worked 35 hours at normal pay, 8 hours at time-and-a-half and some hours at double time. Find the number of hours Jack worked at double pay.

Solution

Let the number of hours worked at double time be x.

$$\begin{aligned}986.1 &= 17.3 \times 35 + 17.3 \times 1.5 \times 8 \\ &\quad + 17.3 \times 2 \times x \\ 986.1 &= 813.1 + 34.6x \\ 34.6x &= 986.1 - 813.1 \\ 34.6x &= 173 \\ x &= 5\end{aligned}$$

∴ Jack worked 5 hours at double time.

FOCUS on ...

1. The question: Asks you to find the number of hours Jack worked at double time.
2. The information: Gives Jack's hourly rate, the total amount he is paid and the hours he worked at normal time and time-and-a-half.
3. Your working: You need to introduce a pronumeral to set up an equation to solve: let the number of hours worked at double time be x. Find the total amount that he is paid for working normal time and time-and-a-half.
   $$\begin{aligned}986.1 &= 17.3 \times 35 + 17.3 \times 1.5 \times 8 \\ &\quad + 17.3 \times 2 \times x \\ 986.1 &= 813.1 + 34.6x \\ 34.6x &= 986.1 - 813.1 \\ 34.6x &= 173 \\ \frac{34.6x}{34.6} &= \frac{173}{34.6} \\ x &= 5\end{aligned}$$
4. Your answer: Make sure you write the correct units: Jack worked 5 hours at double time.

Now try these!

1. In one week a gardener worked 28 hours normal pay, 8 hours at time-and-a-half and 6 hours at double time. What was her total pay if she earns $25.60 per hour?

2. Josie is paid $756 for working 18 hours at normal pay, 4 hours at time-and-a-half and 6 hours at double time. What is her hourly rate?

3. A plumber charges a normal hourly rate of $75 per hour. Last week he worked 42 hours at normal pay, 6 hours on Saturday earning time-and-a-half and 5 hours on Sunday earning double time. How much did he earn?

4. One week Tara earned $813.20 when she worked 18 hours at normal pay, 8 hours at time-and-a-half and some hours at double time. How many hours at double time did she work if her rate of pay is $21.40 per hour?

5. Twins Damian and Jordan work in a retail store and are both paid $13.20 per hour. In one week Damien works 10 hours at normal pay and 6 hours at time-and-a-half, while Jordan works 8 hours at normal pay and 7 hours at double time. Who gets paid the larger amount, and by how much?

6. **CHALLENGE** This week Leonard has worked 12 hours at normal pay and 9 hours at time-and-a-half and earned a total pay of $492.15. Next week he is rostered to work 9 hours normal pay and 3 hours double time. If his pay rate remains the same, what will he earn for next week's work?

Answers page 126

KEY SKILL

2 Financial mathematics: Taxation A

HINTS

Taxable income	Tax on this income
0 – \$18 200	Nil
\$18 201 – \$37 000	19c for each \$1 over \$18 200
\$37 001 – \$80 000	\$3572 plus 32.5c for each \$1 over \$37 000
\$80 001 – \$180 000	\$17 547 plus 37c for each \$1 over \$80 000
\$180 001 and over	\$54 547 plus 45c for each \$1 over \$180 000

- Taxable income = Income − Deductions.
- The Australian Taxation Office (ATO) publishes tax tables which are used to determine the tax payable on a taxable income.
  e.g. What is the tax payable on a taxable income of \$63 500?
  $3572 + 0.325 \times (63\,500 - 37\,000) = 12\,184.5$ ∴ tax of \$12 184.50
- Most workers pay instalments throughout the year and if too much tax has been paid will receive a refund:
  Refund = Tax paid − Tax owing.

Examples

Andy has a gross income of \$85 450 and total deductions of \$3600. Calculate the total tax payable on his taxable income using the tax table above.

Solution

$$\text{Taxable income} = 85\,450 - 3600 = 81\,850$$
$$\text{Tax payable} = 17\,547 + 0.37 \times 1850 = 18\,231.5$$

∴ Andy will pay income tax of \$18 231.50.

FOCUS on …

1. The question: Asks you to find the tax Andy will pay on his taxable income.
2. The information: Gives his gross income, his deductions and the tax table to be used.
3. Your working: You need to find the amount of taxable income by subtracting the deductions from his gross income.
   Taxable income = Gross income − Deductions
   = 85 450 − 3600
   = 81 850
   As Andy's taxable income is in the bracket \$80 001 – \$180 000 you use: \$17 547 plus 37c for each \$1 over \$80 000.
   Tax payable = 17 547 + 0.37 × (81 850 − 80 000)
   = 18 231.5
4. Your answer: Make sure you write the correct units: Andy will pay \$18 231.50 tax.

Jonah pays fortnightly tax of \$590. He has an annual salary of \$72 450 and has total deductions of \$2890. What will be Jonah's tax refund from the ATO?

Solution

$$\text{Total of fortnightly tax paid} = 590 \times 26 = 15\,340$$
$$\text{Taxable income} = 72\,450 - 2890 = 69\,560$$
$$\text{Tax payable} = 3572 + 0.325 \times 32\,560 = 14\,154$$
$$\text{Refund} = 15\,340 - 14\,154 = 1186$$

∴ Jonah will receive a refund of \$1186.

FOCUS on …

1. The question: Asks you to find the amount of Jonah's refund.
2. The information: Gives the amount each fortnight Jonah pays in tax, his annual salary and the amount of deductions.
3. Your working: Find the total tax paid throughout the year by multiplying the amount by 26.
   Total tax paid = 590 × 26 = 15 340
   Find the taxable income by subtraction.
   Taxable income = Gross income − Deductions
   = 72 450 − 2890 = 69 560
   As Jonah's taxable income is in the bracket \$37 001 – \$80 000 you use: \$3572 plus 32.5c for each \$1 over \$37 000.
   Tax payable = 3572 + 0.325 × (69 560 − 37 000)
   = 14 154
   Find the refund by subtraction.
   Refund = Tax paid − Tax owing
   = 15 340 − 14 154 = 1186
4. Your answer: Make sure you write the correct units: Jonah will receive a refund of \$1186.

Now try these!

Use the tax table to answer questions 1 to 6.

Taxable income	Tax on this income
0 – $18 200	Nil
$18 201 – $37 000	19c for each $1 over $18 200
$37 001 – $80 000	$3572 plus 32.5c for each $1 over $37 000
$80 001 – $180 000	$17 547 plus 37c for each $1 over $80 000
$180 001 and over	$54 547 plus 45c for each $1 over $180 000

1. Amelia earns $46 920 and has total deductions of $2390. Find the tax she will pay on her taxable income.

2. Last financial year, Cooper paid $788 per fortnight tax. He earned $86 300 and had total deductions of $2540. How much tax refund will Cooper receive?

3. Mia listed her tax deductions as uniform purchase and cleaning: $670 and child sponsorship: $504. If her gross salary was $72 400, how much money will she pay in tax?

4. Heidi works part-time for two companies. She earned $38 600 in one job and $16 460 in the other. If she paid a total of $9430 in tax, will she receive a refund from the ATO, or will she need to pay more tax?

5. Last financial year Jayden's taxable income was $89 400 p.a. His wife Bella works part-time and her taxable income was exactly half of her husband's. For each of them, calculate the percentage of their taxable income paid in income tax.

6. CHALLENGE Each fortnight Ashley is paid $3200. She has been offered a pay rise of 2.5% p.a. After paying the additional tax, how much more will Ashley receive in her fortnightly pay packet?

Answers page 126

KEY SKILL

3 Financial mathematics: Taxation B

HINTS

- Make sure you know the **Taxation Hints** from Key Skill 2 on page 12.

Taxable income	Tax on this income
0 – \$18 200	Nil
\$18 201 – \$37 000	19c for each \$1 over \$18 200
\$37 001 – \$80 000	\$3572 plus 32.5c for each \$1 over \$37 000
\$80 001 – \$180 000	\$17 547 plus 37c for each \$1 over \$80 000
\$180 001 and over	\$54 547 plus 45c for each \$1 over \$180 000

Examples

Emily paid \$12 113 tax. Using the tax table above, calculate her taxable income.

Solution

Her income was between \$37 000 and \$80 000.
Let her income be x.

$$12\,113 = 3572 + 0.325 \times (x - 37\,000)$$
$$8541 = 0.325(x - 37\,000)$$
$$x - 37\,000 = \frac{8541}{0.325}$$
$$= 26\,280$$
$$x = 63\,280$$

∴ Emily had a taxable income of \$63 280.

FOCUS on ...

1. The question: Asks you to find Emily's taxable income.
2. The information: Gives the amount of tax paid and the tax table.
3. Your working: You need to find the row of the table that is used. Emily's tax of \$12 113 is more than \$3572 and less than \$17 547. This means you use the row for income: \$37 001 – \$80 000.
   You need to introduce a pronumeral to set up an equation to solve: Let Emily's income amount = x.
   $$12\,113 = 3572 + 0.325 \times (x - 37\,000)$$
   $$8541 = 0.325(x - 37\,000)$$
   $$x - 37\,000 = \frac{8541}{0.325} = 26\,280$$
   $$x = 63\,280$$
4. Your answer: Make sure you write the correct units: Emily's taxable income was \$63 280.

Elizabeth received a tax refund of \$1353 from the ATO after paying \$1040 per fortnight throughout the year. If she had total deductions of \$2430, find her gross annual earnings.

Solution

$$\text{Total fortnightly tax paid} = 1040 \times 26$$
$$= 27\,040$$
$$\text{Tax payable from table} = 27\,040 - 1353$$
$$= 25\,687$$

Her income was between \$80 000 and \$180 000.
Let her income be x.

$$25\,687 = 17\,547 + 0.37 \times (x - 80\,000)$$
$$8140 = 0.37(x - 80\,000)$$
$$x - 80\,000 = \frac{8140}{0.37} = 22\,000$$
$$x = 102\,000$$

Gross earnings = 102 000 + 2430 = 104 430
∴ Elizabeth had gross annual earnings of \$104 430.

FOCUS on ...

1. The question: Asks you to find Elizabeth's annual earnings.
2. The information: Gives the tax refund, the fortnightly tax paid and the amount of deductions.
3. Your working: You need to find the total tax paid by multiplying the fortnightly amount by 26.
   $$\text{Tax paid} = 1040 \times 26 = 27\,040$$
   As Elizabeth received a refund you subtract this amount from the amount she has already paid.
   $$\text{Tax according to table} = 27\,040 - 1353 = 25\,687$$
   You need to find the row of the table that is used. Elizabeth's tax of \$25 687 is more than \$17 547 and less than \$54 547. This means you use the row for incomes: \$80 001 – \$180 000.
   You need to introduce a pronumeral to set up an equation to solve: Let Elizabeth's income amount = x.
   $$25\,687 = 17\,547 + 0.37 \times (x - 80\,000)$$
   $$8140 = 0.37(x - 80\,000)$$
   $$x - 80\,000 = \frac{8140}{0.37} = 22\,000$$
   $$x = 102\,000$$
   To find the gross earnings you need to add her deductions.
   Gross earnings = 102 000 + 2430 = 104 430
4. Your answer: Make sure you write the correct units: Elizabeth's gross earnings were \$104 430.

Now try these!

Use the tax table to answer questions 1 to 6.

Taxable income	Tax on this income
0 – \$18 200	Nil
\$18 201 – \$37 000	19c for each \$1 over \$18 200
\$37 001 – \$80 000	\$3572 plus 32.5c for each \$1 over \$37 000
\$80 001 – \$180 000	\$17 547 plus 37c for each \$1 over \$80 000
\$180 001 and over	\$54 547 plus 45c for each \$1 over \$180 000

1. Last year Xavier paid \$22 949 in tax. What was his taxable income?

2. Charlie received a tax refund of \$896.75 from the ATO after paying \$670 per fortnight tax throughout the year. If he had total deductions of \$1600, find his gross annual earnings.

3. Jordan paid \$34 567 in tax. He had tax deductions totalling \$4520. What was his gross income?

4. Last year Braxton paid \$60 per fortnight in tax and received a refund of \$116 from the ATO. What was his taxable income?

5. After paying \$740 each fortnight tax, Brodie still owed the ATO another \$1560. What was her taxable income, to the nearest 1000 dollars?

6. CHALLENGE Last financial year Cyril earned \$82 300. He paid twice as much tax as his brother Cedric. Find Cedric's taxable income, to the nearest cent.

Answers pages 126–127

KEY SKILL

4 Financial mathematics: Simple interest A

HINTS

- Simple interest is calculated using
  $I = Prn$, where
  I = interest
  P = principal (the amount of investment)
  r = interest rate per time period
  n = number of time periods
  e.g. Calculate interest on \$3000 at 4% p.a. for 5 years.
  $3000 \times 0.04 \times 5 = 600 \quad \therefore \$600$
- To change from years to months you multiply by 12. To change a yearly rate to a monthly rate you divide by 12.
  e.g. Find monthly interest rate of 6% p.a.
  As $6 \div 12 = 0.5$, then $0.5\% = 0.005$.
  $\therefore$ 6% p.a. is 0.5% per month, 0.005 as decimal.

Reminder!

- Read the question carefully to identify what needs to be found.
- Re-read the question when you've finished your answer to make sure that the question has been answered and your solution makes sense.

Examples

Yusef invested \$20 000 for 7 years at an interest rate of 4.5% p.a. What amount of simple interest did he receive?

Solution

Subs. $P = 20\,000$, $r = 0.045$, $n = 7$

$$\begin{aligned} I &= Prn \\ &= 20\,000 \times 0.045 \times 7 \\ &= 6300 \end{aligned}$$

$\therefore$ Yusef received simple interest of \$6300.

FOCUS on ...

1. The question: Asks you to find the amount of simple interest (I) received.
2. The information: Gives the principal, the rate of interest and the number of years.
3. Your working: The percentage needs to be changed to a decimal: 4.5% = 0.045.
   You need to substitute the values of P, r and n into the formula $I = Prn$.
   $$\begin{aligned} I &= 20\,000 \times 0.045 \times 7 \\ &= 6300 \end{aligned}$$
4. Your answer: Make sure you write the correct units: Yusef receives \$6300 interest.

Amelia won \$50 000 in a lottery. She decided to invest the money in a building society account which attracted simple interest of 6% p.a., calculated monthly. If she left the money in the account for 2 years, what was the amount of interest she earned?

Solution

$$\begin{aligned} 6\% \text{ p.a.} &= 0.5\% \text{ per month} \\ &= 0.005 \\ 2 \text{ years} &= 24 \text{ months} \end{aligned}$$

Subs. $P = 50\,000$, $r = 0.005$, $n = 24$

$$\begin{aligned} I &= Prn \\ &= 50\,000 \times 0.005 \times 24 \\ &= 6000 \end{aligned}$$

$\therefore$ Amelia's money earned interest of \$6000.

FOCUS on ...

1. The question: Asks you to find the amount of simple interest (I) received.
2. The information: Gives the principal, the rate of interest, the number of years and that the interest is calculated monthly.
3. Your working: The annual percentage rate needs to be converted to a monthly rate and written as a decimal: 6%/year = 0.5%/month. Also, 0.5% = 0.005. The number of months is also calculated: $2 \times 12 = 24$.
   You need to substitute the values of P, r and n into the formula $I = Prn$.
   $$\begin{aligned} I &= 50\,000 \times 0.005 \times 24 \\ &= 6000 \end{aligned}$$
4. Your answer: Make sure you write the correct units: Amelia receives \$6000 interest.

Now try these!

1. Chelsea invested $3800 in an account attracting simple interest of 3.75% p.a. for 4 years. What was the amount of interest earned?

2. Noah invests $35 000 in a special savings account that offers interest of 6.6% p.a. simple interest, calculated monthly. What amount of interest will be earned after 2 years?

3. Caitlin invested $8500 in an account offering 1.6% per quarter for 18 months. How much simple interest will she earn?

4. Ruby is saving for a house and invests $27 000 for 18 months into an account earning 6% p.a. simple interest, calculated monthly. How much interest does Ruby earn?

5. A mother gives her two sons $10 000 each which they invest in different savings accounts. Ewan deposits his money for 5 years at 6.5% p.a. and Oliver invests his money at 5.25% p.a. for 6 years. Which investment earns the most interest, and by how much?

6. CHALLENGE On 1 March 2010, Caden invested $4500 in an account at 3.15% p.a. simple interest. On 1 March 2013, he withdrew his initial investment plus interest and deposited it in another account offering 4.1% p.a. simple interest. He withdrew all of his money on 1 March 2015. What was the total amount of interest earned over the 5 years?

Answers page 127

KEY SKILL

5 Financial mathematics: Simple interest B

HINTS

- Make sure you know the **Simple interest Hints** from Key Skill 4 on page 16.
- To find P, r or n when using $I = Prn$, first substitute known values into the formula, and then change the subject to the unknown pronumeral,

e.g. Find n if $1440 = 8000 \times 0.06 \times n$

$$\begin{aligned} 1440 &= 8000 \times 0.06 \times n \\ 1440 &= 480n \\ 480n &= 1440 \\ n &= \frac{1440}{480} \\ &= 3 \end{aligned}$$

Reminder!

- Read the question carefully to identify what needs to be found.
- Re-read the question when you've finished your answer to make sure that the question has been answered and your solution makes sense.

Examples

Maia invested a sum of money in a bank account for 5 years and earned simple interest of \$262.50. If the investment had earned a rate of $3\frac{1}{2}\%$ p.a., find the amount of Maia's investment.

Solution

Subs $I = 262.5$, $r = 0.035$, $n = 5$

$$\begin{aligned} I &= Prn \\ 262.5 &= P \times 0.035 \times 5 \\ 262.5 &= 0.175P \\ P &= \frac{262.5}{0.175} \\ &= 1500 \end{aligned}$$

∴ Maia invested \$1500.

FOCUS on ...

1. The question: Asks you to find the principal (P) that has been invested.
2. The information: Gives the amount of interest, the rate of interest and the number of years.
3. Your working: The percentage needs to be changed to a decimal: $3\frac{1}{2}\% = 0.035$.
   You need to substitute the values of I, r and n into the formula $I = Prn$.
   $$\begin{aligned} 262.5 &= P \times 0.035 \times 5 \\ &= 0.175P \\ P &= \frac{262.5}{0.175} \\ &= 1500 \end{aligned}$$
4. Your answer: Make sure you write the correct units: Maia invested \$1500.

Charlie earned \$1320 simple interest from an investment of \$12 000 over 4 years. What was the annual interest rate earned on the investment?

Solution

Subs. $I = 1320$, $P = 12\,000$, $n = 4$

$$\begin{aligned} I &= Prn \\ 1320 &= 12\,000 \times r \times 4 \\ 1320 &= 48\,000r \\ r &= \frac{1320}{48\,000} \\ &= 0.0275 \end{aligned}$$

Now, $0.0275 \times 100\% = 2.75\%$.

∴ Charlie invested at 2.75% p.a. interest rate.

FOCUS on ...

1. The question: Asks you to find the annual interest rate on Charlie's investment.
2. The information: Gives the amount of interest, the principal and the number of years.
3. Your working: You need to substitute the values of I, P and n into the formula $I = Prn$.
   $$\begin{aligned} 1320 &= 12\,000 \times r \times 4 \\ &= 48\,000r \\ r &= \frac{1320}{48\,000} \\ &= 0.0275 \end{aligned}$$
   Change to a percentage: $0.0275 \times 100\% = 2.75\%$.
4. Your answer: Make sure you write the correct units: Charlie's interest rate was 2.75% p.a.

Now try these!

1. Morton invested a sum of money at 4% p.a. for 7 years and earned \$1008 simple interest. What was the amount of his original investment?

2. Dalip deposited \$6400 in a savings account for 5 years which earned simple interest of \$1728. What was the annual interest rate on the account?

3. Samuel invested \$8000 in a savings account offering 5% p.a. simple interest. How many years would it take for him to double his money?

4. Grace and Erik invested \$5000 each in separate savings accounts offering the same rate of interest. After 8 years Grace had earned \$1900 simple interest. How much more interest will Erik have earned after another 2 years?

5. Phoebe invests \$9600 in an account offering 6% p.a. simple interest, calculated monthly. For how many years is the money invested if it earns interest of \$1728?

6. CHALLENGE Maxine deposited \$8000 in one account and \$2000 in another. Both investments were for 2 years and earned simple interest. The rate on the first account was 1% lower than the rate on the second account. If the amount of interest was \$320 more on the first account than the second, find the annual rate of interest on both accounts.

Answers pages 127–128

KEY SKILL

6 Financial mathematics: Loans

HINTS

- When an expensive item is purchased the seller may require a deposit and interest will then be charged on the balance owing. The total amount owing is usually paid by monthly instalments.
- A deposit can be an amount, or a percentage, of the cash price, e.g. A deposit of 25% is paid when buying a caravan valued at \$48 000. Find the deposit. $0.25 \times 48\,000 = 12\,000$
  ∴ the deposit will be \$12 000.
- Loan instalments or repayments are usually paid monthly, e.g. A car is to be purchased over 5 years. How many monthly instalments will there be? $12 \times 5 = 60$
  ∴ there will be 60 monthly instalments.

Examples

A lounge suite has a cash price of \$3200 but is to be purchased with a deposit of 25% and interest of 8% p.a. charged on the balance owing over 4 years. Find each monthly instalment.

Solution

$$\begin{aligned}\text{Deposit} &= 0.25 \times 3200 = 800\\ \text{Balance owing} &= 2400\\ \text{Interest} &= 2400 \times 0.08 \times 4\\ &= 768\\ \text{Total repayments} &= 3168\\ \text{Monthly repayments} &= 3168 \div 48\\ &= 66\end{aligned}$$

∴ each monthly instalment is \$66.

FOCUS on ...

1. The question: Asks you to find the monthly instalment.
2. The information: Gives the cash price, the percentage deposit, the interest charged and the length of the loan (which tells you the number of instalments).
3. Your working: You need to find the deposit.
   Deposit = 0.25 × 3200 = 800
   Subtract to find the amount still owing.
   Balance owing = 3200 − 800 = 2400
   Use $I = Prn$ to find the interest.
   $I = 2400 \times 0.08 \times 4 = 768$
   Add the interest to the balance owing.
   Total instalments = 2400 + 768 = 3168
   Divide to find the monthly instalment.
   Monthly instalments = 3168 ÷ 48 = 66
4. Your answer: Make sure you write the correct units:
   The monthly instalment is \$66.

William agrees to pay 48 monthly instalments of \$360 when he purchases a motorbike. If the deposit on the bike is 20% of the original price of \$18 000, what annual interest rate will William be charged?

Solution

$$\begin{aligned}\text{Deposit} &= 0.2 \times 18\,000 = 3600\\ \text{Balance owing} &= 14\,400\\ \text{Total instalments} &= 360 \times 48 = 17\,280\\ \text{Total interest} &= 17\,280 - 14\,400 = 2880\\ I &= Prn\\ 2880 &= 14\,400 \times r \times 4\\ 2880 &= 57\,600r\\ r &= 2880 \div 57\,600 = 0.05\end{aligned}$$

∴ William will be charged interest of 5% p.a.

FOCUS on ...

1. The question: Asks you to find the interest rate.
2. The information: Gives the cash price, the percentage deposit, the length of the loan (number of months) and the monthly instalments.
3. Your working: You need to find the deposit.
   Deposit = 0.2 × 18 000 = 3600. Subtract to find the amount still owing.
   Balance owing = 18 000 − 3600 = 14 400
   Multiply to find the total instalments.
   Total instalments = 360 × 48 = 17 280
   Subtract to find the interest that has been charged.
   Total interest = 17280 − 14 400 = 2880
   Use $I = Prn$ to find the annual interest rate, where the principal = balance owing, and 48 months = 4 years.
   $$2880 = 14\,400 \times r \times 4$$
   $$57\,600r = 2880$$
   $$r = \frac{2880}{57\,600} = 0.05$$
   Convert decimal to percentage: 0.05 = 5%
4. Your answer: Make sure you write the correct units:
   William will be charged 5% p.a.

Now try these!

1. A bedroom suite is priced at $2590 but is to be purchased on the following terms: 20% deposit, 6% p.a. interest and 24 monthly repayments. How much is each monthly repayment?

2. To buy a $54 000 boat Alyssa is charged a deposit of $6000 and pays the balance off over 5 years with monthly instalments of $1040. What is the annual interest rate she is charged, to the nearest per cent?

3. Owen trades in his old car for $5000 and buys a car priced at $52 000 by paying an interest rate of 4.1% p.a. Find the size of each of the 48 monthly repayments.

4. A motor bike has a cash price of $18 400 but is to be purchased by leaving a deposit of $3200 and paying the balance in monthly instalments over 5 years. If each instalment is $304, what is the annual interest rate?

5. Phil bought a $36 000 car from Kel's Autos by paying a 5% deposit. He had two options to pay the balance owing:
Option A: 4% p.a. interest over 3 years.
Option B: 3.2% p.a. interest over 4 years.
What was the difference between monthly repayments for the two options?

6. CHALLENGE Jenny bought a car valued at $40 000 by paying a deposit of 10% and paying the balance with an interest rate of 6% p.a. If her monthly repayments were $930, how many repayments did she make?

Answers page 128

REVISION TEST  Level of difficulty—Average

1. Anne is paid $1160.16 for working 24 hours at normal pay, 8 hours at time-and-a-half and 6 hours at double time. What is her hourly rate?

This table is used to answer questions 2 and 3.

Taxable income	Tax on this income
0 – $18 200	Nil
$18 201 – $37 000	19c for each $1 over $18 200
$37 001 – $80 000	$3572 plus 32.5c for each $1 over $37 000
$80 001 – $180 000	$17 547 plus 37c for each $1 over $80 000
$180 001 and over	$54 547 plus 45c for each $1 over $180 000

2. Last financial year, Rod paid $625 per fortnight tax. He earned $74 940 and had total deductions of $1840. How much tax refund will Rod receive?

3. Last year Lara paid $27 167 in tax. What was her taxable income?

4. Craig invests $20 000 in a special savings account that offers interest of 6% p.a. simple interest, calculated monthly. What amount of interest will be earned after 3 years?

5. Liam invests $9600 in an account offering 4% p.a. simple interest, calculated quarterly. For how many years is the money invested if it earns interest of $1152?

6. A jet-ski has a cash price of $15 600 but is to be purchased by leaving a deposit of 15% and paying the balance in monthly instalments over 4 years. If each instalment is $353.60, what is the annual interest rate?

Answers pages 128–129

REVISION TEST  Level of difficulty—Challenging

1. Last week Nathan worked 28 hours at normal pay and 6 hours at time-and-a-half and was paid $794.02. This week he works 30 hours at normal pay and 6 hours at double time. If his pay rate is unchanged, what will be his total pay this week?

Use the tax table to answer questions 2 and 3.

Taxable income	Tax on this income
0 – $18 200	Nil
$18 201 – $37 000	19c for each $1 over $18 200
$37 001 – $80 000	$3572 plus 32.5c for each $1 over $37 000
$80 001 – $180 000	$17 547 plus 37c for each $1 over $80 000
$180 001 and over	$54 547 plus 45c for each $1 over $180 000

2. Nicole is paid $2350 each fortnight. She is to receive a pay increase of 3% p.a. After paying the additional tax, how much extra will she receive each fortnight?

3. Anita earned $88 400 and paid twice as much tax as Kerri. What was Kerri's taxable income, to the nearest cent?

4. Jade invested $5000 in one account earning 3.45% p.a. simple interest for 2 years. She then withdrew all the money in the account and deposited it in another account earning 3.7% p.a. simple interest for another 2 years. What was the total amount of interest Jade earned over the 4 years?

5. Cosette invested $12 000 at one bank and $8000 at another. Both investments attracted simple interest for 3 years. The rate on the first investment was 0.75% lower than the rate on the second investment. If she earned $600 more on her first investment than on her second, find the annual rate of interest on both accounts.

6. The price of a caravan is $60 000. Bruce purchased the caravan by paying a deposit of 20% and the balance on an interest rate of 4% p.a. If the monthly repayments were $960, how many repayments did Bruce have to make?

Answers page 129

KEY SKILL

7 Financial mathematics: Compound interest A

HINTS

- Compound interest is calculated using $A = P(1 + r)^n$, where
  A = total amount (or compounded amount)
  P = principal (the amount of investment)
  r = interest rate per time period
  n = number of time periods
  e.g. Calculate $2000 \times 1.08^5 = 2939$ (nearest whole)
- The amount of compound interest is found by subtracting the principal from the compounded amount.
- Take note of the regularity of the compounding: annually, quarterly, monthly, daily, etc.,
  e.g. If the annual rate of interest is 18%, find the daily rate as a decimal, correct to 4 decimal places.
  $18 \div 365 = 0.049\,315\,068\ldots$, or 0.0493 (4 dec. pl.)
  $\therefore$ 18% p.a. is 0.0493% per day, or 0.000 493 as a decimal.

Examples

Find the amount of compound interest if Megan invests $12 000 for 5 years at an interest rate of 4.5% p.a.

Solution

Subs. $P = 12\,000$, $r = 0.045$, $n = 5$

$$A = P(1 + r)^n$$
$$A = 12\,000(1.045)^5$$
$$= 14\,954.1832\ldots$$
$$= 14\,954.18 \text{ (2 dec. pl.)}$$
$$\text{Interest} = 14\,954.18 - 12\,000$$
$$= 2954.18$$

$\therefore$ Megan earns interest of $2954.18.

FOCUS on...

1. The question: Asks you to find the amount of compound interest.
2. The information: Gives the principal, the number of years and the interest rate p.a.
3. Your working: You need to substitute into the formula: use $P = 12\,000$, $r = 0.045$, $n = 5$.
   $A = P(1 + r)^n$
   $A = 12\,000(1.045)^5$
   $= 14\,954.18$ (2 dec. pl.)
   Now, subtract the principal from the amount A.
   Interest $= 14\,954.18 - 12\,000$
   $= 2954.18$
4. Your answer: Make sure you write the correct units: Megan earns compound interest of $2954.18.

Pablo invests $6500 in an account that offers 6% p.a., compounded monthly. Find the amount of interest after 3 years.

Solution

Monthly: $6 \div 12 = 0.5$

$\therefore r = 0.5\%$ per month, or 0.005

Also, $n = 3$ years $= 36$ months

Subs. $P = 6500$, $r = 0.005$, $n = 36$

$$A = P(1 + r)^n$$
$$A = 6500(1.005)^{36}$$
$$= 7778.423\,411\ldots$$
$$= 7778.42 \text{ (2 dec. pl.)}$$
$$\text{Interest} = 7778.42 - 6500$$
$$= 1278.42$$

$\therefore$ Pablo earns interest of $1278.42.

FOCUS on...

1. The question: Asks you to find the amount of compound interest.
2. The information: Gives the principal, the interest rate p.a. and the number of years. It also says that the investment is compounded monthly.
3. Your working: First change the interest rate, and time, to months. As $6 \div 12 = 0.5$, then rate is 0.5% per month (or 0.005 as decimal). Also, as $3 \times 12 = 36$, then use $n = 36$. You need to substitute into the formula: use $P = 6500$, $r = 0.005$, $n = 36$.
   $A = P(1 + r)^n$
   $A = 6500(1.005)^{36}$
   $= 7778.42$ (2 dec. pl.)
   Now, subtract the principal from the amount A.
   Interest $= 7778.42 - 6500$
   $= 1278.42$
4. Your answer: Make sure you write the correct units: Pablo earns compound interest of $1278.42.

Now try these!

1. Takumi invested \$7000 in an account offering compound interest of 4.25% p.a. for 10 years. Find the amount of interest.

2. Ingrid is saving for a house and invests \$28 000 in a savings account for 18 months. The account offers compound interest of 4.8% p.a., compounded monthly. What amount of interest does she earn on her investment?

3. The price of an apartment was \$462 000 in January 2011. What was the price in January 2015 if the value rose by 3.1% annually? Give your answer to the nearest dollar.

4. Magnus has \$20 000 to invest for 3 years. He deposits half the amount in an account offering 5% p.a. compounded annually and the remainder in another account offering 4% p.a. compounded quarterly. Which account will gain more interest and by what amount?

5. Two years ago the population of a village in Somalia was 24 000. If the birth rate was 4.2% p.a. and the death rate was 1.8% p.a., what is the present population?

6. CHALLENGE Oscar borrowed \$5000 from Lukas, paying him 6% p.a. simple interest. He then lent the money to Kate and charged her 7.2% p.a. interest, compounded monthly. If the loans lasted a period of 8 years, how much money did Oscar make on the loans?

Answers pages 129–130

KEY SKILL

8 Financial mathematics: Compound interest B

HINTS

- Make sure you know the **Compound interest Hints** from Key Skill 7 on page 24.
- To find P, R or n when using $A = P(1 + r)^n$, first substitute known values into the formula, and then change the subject to the unknown pronumeral.

Reminder!

- Read the question carefully to identify what needs to be found.
- Re-read the question when you've finished your answer to make sure that the question has been answered and your solution makes sense.

Examples

Ali invested a sum of money earning an interest rate of 6.4% p.a. compounded annually. At the end of 6 years the investment had grown to \$14 510. Find Ali's original investment, to the nearest dollar.

Solution

Subs. $A = 14\,510$, $r = 0.064$, $n = 6$

$$A = P(1 + r)^n$$
$$14\,510 = P(1.064)^6$$
$$P = \frac{14\,510}{1.064^6}$$
$$= 10\,000.406\,29\ldots$$
$$= 10\,000 \text{ (nearest whole)}$$

$\therefore$ Ali originally invested \$10 000.

FOCUS on ...

1. The question: Asks you to find the original investment (principal).
2. The information: Gives the interest rate, the number of years and the compounded amount.
3. Your working: You need to change the interest rate to a decimal: 6.4% = 0.064. You need to substitute into the formula: use $A = 14\,510$, $r = 0.064$, $n = 6$.
$$A = P(1 + r)^n$$
$$14\,510 = P(1.064)^6$$
$$P = \frac{14\,510}{1.064^6}$$
$$= 10\,000.406\,29\ldots$$
$$= 10\,000 \text{ (nearest whole)}$$
4. Your answer: Make sure you write the correct units: Ali invested \$10 000.

Savannah invested \$8000 for 3 years. If she earned \$1000 compound interest, find the annual interest rate, to the nearest whole.

Solution

$8000 + 1000 = 9000$

Subs. $A = 9000$, $P = 8000$, $n = 3$

$$A = P(1 + r)^n$$
$$9000 = 8000(1 + r)^3$$
$$(1 + r)^3 = 1.125$$
$$1 + r = \sqrt[3]{1.125}$$
$$r = \sqrt[3]{1.125} - 1$$
$$= 0.040\,041\,911\ldots$$
$$= 0.04 \text{ (2 dec. pl.)}$$

$\therefore$ Savannah's investment earned 4% p.a.

FOCUS on ...

1. The question: Asks you to find the annual interest rate.
2. The information: Gives the original investment (principal), the number of years and the amount of compound interest.
3. Your working: You need to add the principal to the compound interest to find the compounded amount: $A = 8000 + 1000 = 9000$
You need to substitute into the formula: use $A = 9000$, $P = 8000$, $n = 3$.
$$A = P(1 + r)^n$$
$$9000 = 8000(1 + r)^3$$
$$(1 + r)^3 = \frac{9000}{8000}$$
$$(1 + r)^3 = 1.125$$
$$1 + r = \sqrt[3]{1.125}$$
$$r = \sqrt[3]{1.125} - 1$$
$$= 0.040\,041\,911\ldots$$
$$= 0.04 \text{ (2 dec. pl.)}$$
Convert decimal to percentage: 0.04 = 4%
4. Your answer: Make sure you write the correct units: Savannah was charged 4% p.a.

Now try these!

1. Penelope invested a sum of money in an account offering 8.5% p.a. compound interest. After 4 years the investment had grown to $13 027. To the nearest dollar, what was Penelope's original investment?

2. Emily deposited $26 000 in an account for 10 years and earned interest of $21 672. What was the annual rate of compound interest on the account? Give your answer as a percentage, correct to two decimal places.

3. Ray borrowed an amount of money for 3 years and was charged 5% p.a. compounded annually. If he repaid a total of $3000, what was the original amount borrowed? Leave your answer to the nearest dollar.

4. Julia deposited $9500 in an account offering an interest rate that compounded quarterly. After 3 years she had earned $1527 interest. What was the annual interest rate on her investment, to the nearest whole?

5. The population of a city consistently rises by 0.4% per month. At the start of November 2014, the population was 28 000. What had been the population at the start of June 2014, and give an estimate for the population in November 2020? Give your answers to the nearest thousand.

6. CHALLENGE Martina invested $5000 in a credit union account offering 3% p.a. simple interest. On the same day, her friend Elena invested a sum of money in a building society account offering 3% p.a. compound interest. After 5 years both girls had earned the same amount of interest. How much did Elena invest, to the nearest dollar?

Answers page 130

KEY SKILL

9 Financial mathematics: Compound interest C

HINTS

- Make sure you know the **Compound interest Hints** from Key Skills 7 and 8 on pages 24 and 26.
- When you use a 'trial and error' or 'guess and refine' method, make sure you show some evidence in your working,

e.g. Use your calculator to estimate x, if $5^x = 10$. Give your answer correct to one decimal place.

Try $x = 1.3$: $\therefore 5^{1.3} = 8.10\ldots$

Try $x = 1.4$: $\therefore 5^{1.4} = 9.51\ldots$

Try $x = 1.5$: $\therefore 5^{1.5} = 11.18\ldots$

$\therefore x = 1.4$ (1 dec. pl.)

Examples

Destiny was given \$8000 when her grandmother passed away and she invested it at an interest rate of 4% p.a. After a number of years the investment had grown to \$9733. To the nearest year, what was the length of her investment?

Solution

Subs $A = 9733$, $P = 8000$, $r = 0.04$

$$A = P(1 + r)^n$$

$$9733 = 8000(1.04)^n$$

$$1.04^n = \frac{9733}{8000} = 1.216\,625$$

Try $n = 3$: $\therefore 1.04^3 = 1.1248\ldots$

Try $n = 5$: $\therefore 1.04^5 = 1.2166\ldots$

$\therefore n = 5$ (nearest whole)

$\therefore$ Destiny invested for 5 years.

FOCUS on ...

1. The question: Asks you to find the number of years of the investment.
2. The information: Gives the principal, the interest rate and the compounded amount.
3. Your working: You need to substitute into the formula: use $A = 9733$, $P = 8000$, $r = 0.04$.

   $$A = P(1 + r)^n$$

   $$9733 = 8000(1.04)^n$$

   $$1.04^n = \frac{9733}{8000} = 1.216\,625$$

   Using your calculator, guess different values of n, where n is a whole number:

   Use $n = 3$: $1.04^3 = 1.124$ (too low)

   Use $n = 4$: $1.04^4 = 1.1698$ (too low)

   Use $n = 5$: $1.04^5 = 1.2166$ (close)

   Use $n = 6$: $1.04^6 = 1.2653$ (too high)

   The best result was $n = 5$.
4. Your answer: Make sure you write the correct units: Destiny invested for 5 years.

The population in Greaves in January 1990 was 7820. Each year the population grew at a rate of 2.5%. In what year will the population exceed 20 000?

Solution

Subs. $A = 20\,000$, $P = 7820$, $r = 0.025$

$$A = P(1 + r)^n$$

$$20\,000 = 7820(1.025)^n$$

$$1.025^n = \frac{20\,000}{7820} = 2.557\,5447\ldots$$

Try $n = 30$: $\therefore 1.025^{30} = 2.0975\ldots$

Try $n = 40$: $\therefore 1.025^{40} = 2.6850\ldots$

Try $n = 38$: $\therefore 1.025^{38} = 2.5556\ldots$

$1990 + 38 = 2028$

$\therefore$ the town will exceed 20 000 in 2028.

FOCUS on ...

1. The question: Asks you to find the year the population will pass 20 000.
2. The information: Gives the existing population (principal), the rate of increase and the future population (compounded amount).
3. Your working: You need to substitute into the formula: use $A = 20\,000$, $P = 7820$, $r = 0.025$.

   $$A = P(1 + r)^n$$

   $$20\,000 = 7820(1.025)^n$$

   $$1.025^n = \frac{20\,000}{7820} = 2.557\,5447\,57\ldots$$

   Using your calculator, guess different values of n, where n is a whole number:

   Use $n = 30$: $1.025^{30} = 2.098$ (too low)

   Use $n = 40$: $1.025^{40} = 2.685$ (too high)

   Use $n = 35$: $1.025^{35} = 2.373$ (closer)

   Use $n = 38$: $1.025^{38} = 2.5556$ (very close)

   The best result was $n = 38$.

   Add to find the year: $1990 + 38 = 2028$.
4. Your answer: Make sure you write the correct units: The population will exceed 20 000 in 2028.

Now try these!

1 Ashley invested $14 000 in a credit union account offering an interest rate of 6.25% p.a., compounded annually. When Ashley decided to withdraw the money, the amount in the account had grown to $22 738. For how many years did Ashley have the money invested? Give your answer to the nearest year.

2 The population of Huntlee was 740 and was increasing at the rate of 0.6% per month. After how much time will the population reach 1000? Give your answer to the nearest month.

3 A painting appreciates in value each year by 2%. If the painting is presently valued at $105 000, after how many years will its value be doubled?

4 On 4 June, the number of confirmed cases of a virus in a town is 16. If the number of cases increases by 6% each day, in what month will the number of confirmed cases reach 1000?

5 Katia was born on 1 May 2010 and on that day her father invested $5000 in a savings account at a fixed interest rate of 4.8% p.a., compounded annually. How old will Katia be when the investment is worth $100 000?

6 CHALLENGE Abigail invests $20 000 in an account offering 0.5% per month simple interest. On the same day Keith invests the same amount of money in another account offering 0.5% per month compound interest. How much more quickly than Abigail will Keith earn $5000 interest?

Answers pages 130–131

KEY SKILL

10 Financial mathematics: Depreciation

HINTS

- Depreciation is the loss of value of an item over time, such as a car, a computer, etc. and is calculated using a variation of the compound interest formula:
  $A = P(1 - r)^n$, where
  A = new value of the item (often called salvage value)
  P = principal (the original value of the item)
  r = interest rate per time period
  n = number of time periods,
  e.g. Calculate
  $5000 \times 0.93^5 = 3478$ (nearest whole)
- To find P, r or n when using $A = P(1 - r)^n$, first substitute known values into the formula, and then change the subject to the unknown pronumeral.

Examples

Elijah purchased a car in January 2015 for \$52 000. If the car depreciates in value at the rate of 18% p.a., find its value in January 2020, to the nearest dollar.

Solution

Subs. $P = 52\,000$, $r = 0.18$, $n = 5$

$$\begin{aligned} A &= P(1 - r)^n \\ &= 52\,000(0.82)^5 \\ &= 19\,278 \text{ (nearest whole)} \end{aligned}$$

∴ Elijah's car is valued at \$19 278.

FOCUS on...

1. The question: Asks you to find the value of the car.
2. The information: Gives the original value (principal), the time period (2015 to 2020) and the rate of depreciation per annum.
3. Your working: You need to subtract to find the number of years: 2020 − 2015 = 5.
   You need to substitute into the formula: use $P = 52\,000$, $r = 0.18$, $n = 5$.
   $$\begin{aligned} A &= P(1 - r)^n \\ A &= 52\,000(0.82)^5 \\ &= 19\,278.47185\ldots \\ &= 19\,278 \text{ (nearest whole)} \end{aligned}$$
4. Your answer: Make sure you write the correct units: The car is worth \$19 278 in 2020.

The company's share price dropped 1% each week for 20 weeks to a price of \$3.20 per share. Find the original price of each share, to the nearest cent, and how much George lost on his portfolio of 1600 shares?

Solution

Subs. $A = 3.2$, $r = 0.01$, $n = 20$

$$\begin{aligned} A &= P(1 - r)^n \\ 3.2 &= P(0.99)^{20} \\ P &= \frac{3.2}{0.99^{20}} \\ &= 3.912\,425\,55\ldots \\ &= 3.91 \text{ (2 dec. pl.)} \end{aligned}$$

∴ the price was \$3.91 per share.

$$\begin{aligned} \text{Loss per share} &= 3.91 - 3.2 \\ &= 0.71 \\ \text{George's total loss} &= 0.71 \times 1600 \\ &= 1136 \end{aligned}$$

∴ George lost \$1136.

FOCUS on...

1. The question: Asks you to find the original price of a share, and the amount of George's loss.
2. The information: Gives the eventual value (A), the number of weeks, the percentage drop each week and the number of shares owned by George.
3. Your working: You need to substitute into the formula: use $A = 3.2$, $r = 0.01$, $n = 20$.
   $$\begin{aligned} A &= P(1 - r)^n \\ 3.2 &= P(0.99)^{20} \\ P &= \frac{3.2}{0.99^{20}} \\ &= 3.912\,425\,55\ldots \\ &= 3.91 \text{ (2 dec. pl.)} \end{aligned}$$
   The original price was \$3.91 per share.
   You subtract to find the loss per share:
   Loss = 3.91 − 3.2 = 0.71
   Mulitply to find George's total loss:
   Total loss = 0.71 × 1600
   = 1136
4. Your answer: Make sure you write the correct units: George lost \$1136.

Now try these!

1. A caravan was purchased in September 2011 for $62 000. If it depreciated annually at the rate of 15%, what was the value of the caravan in September 2014?

2. Sergie purchased 1500 shares in a company. Each day the price of the shares dropped by 1.5%. After 10 trading days the shares were priced at $5.24 each. What did Sergie lose on his portfolio of shares over the period, to the nearest 100 dollars?

3. The original price of a tractor was $82 000. If it depreciates by 18% per year, what will be the value of the tractor after 15 years, to the nearest 100 dollars?

4. In March 2009, Jack bought a motorhome for $92 000. By March 2013, the motorhome had depreciated in price to $45 804. What was the annual rate of depreciation?

5. On 1 May, the Australian dollar was worth 0.68 euros. If it dropped 1.2% each month, how many euros could be exchanged for $A2000 on 1 October? Give your answer to the nearest ten euros.

6. CHALLENGE From 2006 to 2012, the value of a holiday unit appreciated in price by 7% p.a. From 2012 to 2014, the price dropped by 12% p.a. What was the value of the holiday unit in 2014, if it was valued at $240 000 in 2006? Give your answer to the nearest thousand.

Answers page 131

REVISION TEST 3 Level of difficulty—Average

1 Ruby invested \$6400 in an account offering compound interest of 3.75% p.a. for 6 years. Find the amount of interest.

2 The price of a painting was \$820 000 in April 2007. What was the price in April 2015 if the value of the painting rose by 4.3% annually? Give your answer to the nearest 1000 dollars.

3 Bethany deposited \$14 400 in an account for 8 years and earned interest of \$6078. What was the annual rate of compound interest on the account? Give your answer as a percentage, correct to one decimal place.

4 A signed cricket bat appreciates in value by 4% each year. If the bat is presently valued at \$16 400, after how many years will its value be doubled?

5 A builder bought a new work vehicle for \$38 000. If it depreciates by 26% per year, what will be the value of the vehicle after 6 years, to the nearest 100 dollars?

6 Trevor purchased a caravan in May 2009 for \$78 000. By May 2014 the caravan was worth \$51 175.80. What was the annual rate of depreciation on the caravan?

Answers pages 131–132

REVISION TEST  Level of difficulty—Challenging

1. The population of a city consistently rises by 0.36% per month. At the start of April 2013, the population was 25 200. What had been the population at the start of April 2012, and give an estimate for the population in October 2018? Give your answers to the nearest hundred.

2. Natalia deposited $14 000 in an account offering an interest rate that compounded quarterly. After 5 years she had earned $1164 interest. What was the annual interest rate on her investment, to the nearest whole?

3. Gemma borrowed $16 000 from Adam paying him 5% p.a. simple interest. She then lent the money to Magda and charged her 7.8% p.a. interest, compounded monthly. If the loans lasted a period of 6 years, how much money did Gemma make on the loans?

4. Emma invests $10 000 in an account offering 6% p.a., compounded monthly. On the same day her brother Gavin invests $8000 in another account offering 7.2% p.a., compounded monthly. By the time Emma has earned $4000 interest, how much interest has Gavin earned? Give your answer to the nearest dollar.

5. A car depreciates in value by 18% p.a. in the first 5 years after purchase and then at the rate of 12% p.a. over the following 5 years. Find the value of the car after 10 years if it was purchased for $32 000, to the nearest dollar.

6. From 2002 to 2010, the value of a holiday unit appreciated in price by 4.6% p.a. From 2010 to 2014, the price dropped by 5.3% p.a. What was the value of the holiday unit in 2014, if it was valued at $310 000 in 2002? Give your answer to the nearest thousand.

Answers page 132

KEY SKILL

11 Using algebra A

HINTS

- An algebraic expression is a combination of numbers and pronumerals (or variables).
- Check to see that your answer is correct by substituting numbers into the expression,
  e.g. The length of a rectangle is 3 units longer than its width. What is the perimeter of the rectangle if the length is t units?

$$\begin{aligned} \text{length} &= t,\ \text{width} = t - 3 \\ \text{Perimeter} &= 2(t + t - 3) \\ &= 2(2t - 3) \end{aligned}$$

Reminder!

- Read the question carefully to identify what needs to be found.
- Re-read the question when you've finished your answer to make sure that the question has been answered and your solution makes sense.

Examples

Jack is x years old. Aria is 4 years older than 3 times Jack's age 6 years ago. Write an expression for Aria's age in 3 years' time.

Solution

Jack's age 6 years ago $= x - 6$

'3 times': $3(x - 6)$

'4 more': $3(x - 6) + 4$

'In 3 years' time': $3(x - 6) + 4 + 3$

$= 3x - 18 + 7$

$= 3x - 11$

$\therefore$ Aria will be $(3x - 11)$ years old.

FOCUS on ...

1. The question: Asks you to write an expression for Aria's age in 3 years' time.
2. The information: Gives you Jack's age now and a relationship between Jack's age and Aria's age.
3. Your working: Subtract to find Jack's age 6 years ago: $x - 6$.
   Multiply to find 3 times this age: $3(x - 6)$.
   Add to find 4 years older than this age: $3(x - 6) + 4$.
   Add 3 to find the age in 3 years' time:
   $$\begin{aligned} 3(x - 6) + 4 + 3 &= 3x - 18 + 4 + 3 \\ &= 3x - 11 \end{aligned}$$
4. Your answer: Make sure you write the correct units: Aria will be $(3x - 11)$ years old.

Lucia had a total of 30 coins in her money jar. She had x 20-cent coins and the remainder were 50-cent coins. What was the total value of the coins, in dollars?

Solution

There were $(30 - x)$ 50-cent coins.

$$\begin{aligned} \text{Total value} &= 20 \times x + 50 \times (30 - x) \\ &= 20x + 1500 - 50x \\ &= 1500 - 30x \end{aligned}$$

$$\begin{aligned} \therefore (1500 - 30x) \text{ cents} &= \$(\frac{1500 - 30x}{100}) \\ &= \$(\frac{150 - 3x}{10}) \end{aligned}$$

$\therefore$ Lucia had $\$(\frac{150 - 3x}{10})$ in her money jar.

FOCUS on ...

1. The question: Asks you to find the total value, in dollars, of coins in a jar.
2. The information: Gives you the total number of 50- and 20-cent coins, and the number of 20-cent coins.
3. Your working: You subtract to find the number of 50-cent coins: $30 - x$.
   Multiply the number of 20-cent coins by 20 and the number of 50-cent coins by 50.
   $$\begin{aligned} \text{Value in cents} &= 20 \times x + 50 \times (30 - x) \\ &= 20x + 1500 - 50x \\ &= 1500 - 30x \end{aligned}$$
   Divide by 100 to change to dollars.
   $$\begin{aligned} \text{Value in dollars} &= \frac{1500 - 30x}{100} \\ &= \frac{10(150 - 3x)}{100} \\ &= \frac{150 - 3x}{10} \end{aligned}$$
4. Your answer: Make sure you write the correct units: Lucia has $\$\frac{150 - 3x}{10}$ in the jar.

Now try these!

1. Lincoln is x years old now. His brother Levi is twice as old as Lincoln was 3 years ago. Write an expression for the sum of their ages in 5 years' time.

2. Oliver had a total of 40 coins in a money jar. He had x 10-cent coins and the remainder were 5-cent coins. Write an expression, in dollars, to represent the total value of the coins.

3. Jake bought p bags of green marbles for x dollars per bag and q bags of red marbles for $2x$ dollars per bag. Write an expression in terms of x to represent the average cost of a bag of marbles.

4. Two hundred tickets were sold for a concert, with adult tickets being \$10 each and children half-price. If there were x adult tickets sold, write an expression for the total value of ticket sales.

5. Arnie bought m boxes of golf balls for \$$p$ and n boxes of golf balls for \$$q$. He later sold all the balls for \$$r$ each. Write an expression for the average profit on each box of balls.

6. CHALLENGE On Sunday Ruby travelled 240 km to her cousin's house. She travelled the first x km at an average speed of 60 km/h, stopped for y minutes and then the remaining distance at an average speed of 80 km/h. Write an expression for the amount of time it took for the entire journey.

Answers pages 132–133

KEY SKILL

12 Using algebra B

HINTS

- Make sure you know the **Algebra Hints** from Key Skill 11 on page 34.

Examples

Brae works in a coffee shop and is paid $\$x$/hour. On Sunday he worked $3y$ hours at normal pay and y hours at double time. Write an expression for the fraction of Brae's pay that he spent if he saved $\$y$. (Leave your answer as a simplified fraction.)

Solution

Brae worked $3y$ h at normal pay and y h at double time.

$$\text{Total pay} = 3y \times x + 2 \times y \times x = 5xy$$

$$\text{Amount spent} = 5xy - y$$

$$\text{Fraction spent} = \frac{5xy - y}{5xy} = \frac{y(5x - 1)}{5xy} = \frac{5x - 1}{5x}$$

$\therefore$ Brae spent $\frac{5x - 1}{5x}$ of his total pay.

FOCUS on...

1. The question: Asks you to write an expression for the fraction of his pay Brae has spent.
2. The information: Gives you his hourly rate, the hours he worked at different rates and the amount he saved.
3. Your working: You need to multiply and add to find Brae's total pay.
   $3y \times x + 2 \times y \times x = 5xy$
   Subtract the amount he saved: $5xy - y$.
   Write as a fraction:
   $$\frac{5xy - y}{5xy} = \frac{y(5x - 1)}{5xy} = \frac{5x - 1}{5x}$$
4. Your answer: Brae spent $\frac{5x - 1}{5x}$ of his pay.

Serena sold her car and invested a total of $6000 in two bank accounts. She placed $\$x$ in one account earning 4% p.a. interest and the remainder in another account earning 2% p.a. interest. Write an expression for the total amount of simple interest Serena will receive after 18 months. (Leave your answer as a simplified fraction.)

Solution

Serena has $\$x$ earning 4% p.a. interest and $\$(6000 - x)$ earning 2% p.a. interest. Use $I = Prn$.

Total interest

$$= x \times \frac{4}{100} \times \frac{3}{2} + (6000 - x) \times \frac{2}{100} \times \frac{3}{2}$$

$$= \frac{6x}{100} + \frac{3}{100}(6000 - x)$$

$$= \frac{6x}{100} + \frac{18000 - 3x}{100}$$

$$= \frac{3x + 18\,000}{100}$$

$\therefore$ Serena will receive interest of $\$\frac{3x + 18\,000}{100}$.

FOCUS on...

1. The question: Asks you to write an expression for the amount of simple interest.
2. The information: Gives you the total amount invested and the interest rates for two accounts.
3. Your working: You need to subtract from 6000 to find the second investment: $6000 - x$.
   The number of months is changed to years: 18 months $= \frac{3}{2}$ years.
   Use the formula $I = Prn$, and write each interest rate as a fraction.
   $$I = x \times \frac{4}{100} \times \frac{3}{2} + (6000 - x) \times \frac{2}{100} \times \frac{3}{2}$$
   $$= \frac{6x}{100} + \frac{3}{100}(6000 - x)$$
   $$= \frac{6x}{100} + \frac{18\,000 - 3x}{100}$$
   $$= \frac{3x + 18\,000}{100}$$
4. Your answer: Make sure you write the correct units:
   Serena will earn interest of $\$\frac{3x + 18\,000}{100}$.

Now try these!

1. Craig is paid $\$y$ per hour. This week he worked $3x$ hours at normal pay and $2x$ hours at double time. After he was paid he spent $\$5y$ and saved the remainder. Write an expression to represent the fraction of his wage that was saved.

2. Gaby decided to invest a total of \$20 000 in two accounts offering simple interest. She deposited $\$x$ in an account at 5% p.a. interest and the remainder in another account at 4% p.a. interest. Find an expression for the amount of interest earned after 2 years.

3. In a class of 30 students there are x students who are aged y years and the remainder are a year older. What is the average age of the students in the class?

4. The perimeter of a rectangular paddock is $2b$ metres, and its length is a metres. Find an expression for the cost of fertilising the paddock in dollars, if the fertilising costs c cents/m^2.

5. Ginnifer is presently earning $\$x$ per hour for her part-time job. Next week she is to receive a pay rise of p%. What will her total pay be next week if she works m hours normal pay and m hours at time-and-a-half?

6. CHALLENGE Josh won $\$x$ in a lottery and spent \$12 000 of his winnings on an overseas trip. The remainder was invested in an account offering y% p.a. interest, compounded monthly. How much money was in the account after a year?

Answers page 133

KEY SKILL

13 Equations A

HINTS

- An equation is an algebraic expression with an equals sign.
- When solving an equation, the aim is to finish with the pronumeral equal to a number.
- As you solve you must remember to add/subtract/multiply/divide both sides of the equation by the same number at the same time.
- To solve a problem using an equation, let the unknown be x.

Examples

For her 15th birthday party Kristin invited 36 people. If there were 8 more girls than boys, solve an equation to find the number of girls invited to her party.

Solution

Let the number of boys be x.

$\therefore$ the number of girls is $x + 8$.

$$\begin{aligned} x + x + 8 &= 36 \\ 2x + 8 &= 36 \\ 2x &= 36 - 8 \\ 2x &= 28 \\ x &= 14 \\ \text{Now, } x + 8 &= 14 + 8 \\ &= 22 \end{aligned}$$

$\therefore$ Kristin invited 22 girls to her party.

FOCUS on...

1. The question: Asks you to find the number of girls invited.
2. The information: Gives you the total number of people invited and the difference between the number of boys and girls invited.
3. Your working: You need to introduce a pronumeral to set up an equation to solve.
   Let the number of boys = x.
   Add to find the number of girls: x + 8.
   The total of these expressions equals 36.
   $$\begin{aligned} x + x + 8 &= 36 \\ 2x + 8 &= 36 \\ 2x &= 36 - 8 \\ 2x &= 28 \\ x &= 14 \end{aligned}$$
   As girls is x + 8, then 14 + 8 = 22.
4. Your answer: Make sure you write the correct units:
   There were 22 girls invited.

In a match a netball team scored 26 goals. The Goal Shooter (GS) scored 10 more goals than the Goal Attack (GA). Only the GA and GS scored goals. How many goals were scored by each player?

Solution

Let the number of goals scored by the GA be x.

$\therefore$ the number of goals scored by the GS is $x + 10$.

$$\begin{aligned} x + x + 10 &= 26 \\ 2x + 10 &= 26 \\ 2x &= 26 - 10 \\ 2x &= 16 \\ x &= 8 \\ \text{Now, } x + 10 &= 8 + 10 \\ &= 18 \end{aligned}$$

$\therefore$ the GA scored 8 and the GS scored 18.

FOCUS on...

1. The question: Asks you to find the number of goals scored by each player.
2. The information: Gives you the total number of goals scored and the difference between the goals scored by each player.
3. Your working: You need to introduce a pronumeral to set up an equation to solve.
   Let the number of goals scored by GA = x.
   Add to find the number of goals scored by GS: x + 10.
   The total of these expressions equals 26.
   $$\begin{aligned} x + x + 10 &= 26 \\ 2x + 10 &= 26 \\ 2x &= 26 - 10 \\ 2x &= 16 \\ x &= 8 \end{aligned}$$
   As GS is x + 10, then 8 + 10 = 18.
4. Your answer: The GA scored 8 and the GS scored 18.

Now try these!

1. Two cake stalls at the school market sold a total of 96 muffins. If one stall sold 18 more muffins than the other, what was the number sold by each stall?

2. A 2-metre length of ribbon is cut into two pieces. One length is 80 cm shorter than the other. What are the lengths of the two pieces?

3. Glen buys a 5-kg bag of lawn fertiliser and uses 1.2 kg more fertiliser on the backyard lawn than on the lawn in his front yard. How much fertiliser does he use on the front lawn?

4. Kenny walked a certain distance and then ran 2 km further than he walked. He then hitched a ride in a car 4 km further than he had run. If his total distance was 14 km, how far did Kenny travel in the car?

5. Ehmi takes a weekend to drive the 940 km to Adelaide. She drives 80 km further on the first day than on the second day. How far does she drive on the Sunday?

6. CHALLENGE Fiona is paid $14 per hour for her part-time job. In one week she worked 16 hours which was a combination of normal time and double time. If she worked 3 hours more at double time than at normal time, how much was she paid for her work?

Answers pages 133–134

KEY SKILL

14 Equations B

HINTS

- Make sure you know the **Equations Hints** from Key Skill 13 on page 38.

Reminder!

- Read the question carefully to identify what needs to be found.
- Re-read the question when you've finished your answer to make sure that the question has been answered and your solution makes sense.

Examples

On social media, Sienna has half as many friends as Lucas, and Brodie has 3 times as many friends as Sienna. Altogether they have a total of 180 friends. How many friends does each person have?

Solution

Let x be the number of Sienna's friends.

Lucas has $2x$ and Brodie has $3x$.

$$x + 2x + 3x = 180$$
$$6x = 180$$
$$x = 30$$

$\therefore$ Sienna has 30 friends, Lucas has 60 friends and Brodie has 90 friends.

FOCUS on ...

1. The question: Asks you to find the number of social media friends for each person.
2. The information: Gives you the relationship between the number of friends for each person and the total number of friends.
3. Your working: You need to introduce a pronumeral to set up an equation to solve.
   Let Sienna's number of friends = x.
   Multiply to find Lucas's friends: $2 \times x = 2x$.
   Multiply to find Brodie's friends: $3 \times x = 3x$.
   The total number of friends is 180.
   $$x + 2x + 3x = 180$$
   $$6x = 180$$
   $$x = 30$$
   For 2x, then $2 \times 30 = 60$, and $3x = 3 \times 30 = 90$.
4. Your answer: Make sure you write the correct units: Sienna has 30 friends, Lucas has 60 friends and Brodie has 90 friends.

Today, Sebastian is 4 times as old as Brian. In 5 years' time Brian will be half the age Sebastian was 20 years ago. How old will Brian be when Sebastian is 80 years old?

Solution

Let Brian's age now $= x$.

$\therefore$ Sebastian's age now $= 4x$.

In 5 years: Brian will be $x + 5$.

20 years ago: Sebastian was $4x - 20$

$$x + 5 = \frac{1}{2}(4x - 20)$$
$$x + 5 = 2x - 10$$
$$2x - x = 5 + 10$$
$$x = 15$$

Today, Brian is 15 and Sebastian is 60.

As $80 - 60 = 20$, then in 20 years' time Brian will be 35.

$\therefore$ Brian will be 35 years old.

FOCUS on ...

1. The question: Asks you to find the age of Brian when Sebastian is 80 years old.
2. The information: Gives you the relationship between the ages of Sebastian and Brian now, and in 5 years.
3. Your working: You need to introduce a pronumeral to set up an equation to solve.
   Let Brian's age now = x.
   Multiply to find Sebastian's age now: $4 \times x = 4x$.
   Add to find Brian's age in 5 years: $x + 5$.
   Subtract to find Sebastian's age 20 years ago: $4x - 20$.
   $$x + 5 = \frac{1}{2}(4x - 20)$$
   $$x + 5 = 2x - 10$$
   $$2x - x = 5 + 10$$
   $$x = 15$$
   Presently, Brian is 15 and Sebastian is $4 \times 15 = 60$.
   As $80 - 60 = 20$, then in 20 years Brian will be $15 + 20 = 35$.
4. Your answer: Make sure you write the correct units: Brian will be 35 years old.

Now try these!

1. A bag contains twice as many red balls as green balls and four times as many blue balls as green balls. If there are 56 balls in the bag, how many of each colour is in the bag?

2. Today Jack is 5 times the age of his cousin Carmen. In 12 years' time, Carmen will be half Jack's age. How old will Jack be when Carmen is 50 years old?

3. The larger of two numbers is 3 times the smaller number. If the larger number is increased by 12 it is 30 more than the smaller number. What are the two numbers?

4. Today, Jacob is 4 years older than Anthony. Nine years ago Christopher was 3 times as old as Anthony. If the total of the boys' ages is 66, how old will Jacob and Anthony be at Christopher's 30th birthday party?

5. In a game, Zoe has twice as many points as Xavier and Hannah has one-third as many points as Xavier. The three people have a total of 720 000 points. How many points does each player have?

6. CHALLENGE There were 60 people invited to a party. The host ordered enough pizzas so that a whole pizza would be eaten by 2 males or 3 females. If 26 pizzas were ordered, how many females were at the party?

Answers page 134

KEY SKILL

15 Equations and measurement

HINTS

- Make sure you know the **Equations Hints** from Key Skill 13 on page 38.

Examples

If fencing costs \$7.80 per metre, the cost of fencing a triangular paddock is \$2340. The second side is 45 m longer than the first side and the third side is 5 m shorter than twice the length of the first side. What are the lengths of each side of the paddock?

Solution

Perimeter $= 2340 \div 7.8 = 300$

Let the length of the first side be x.

Let the length of the second side be $x + 45$.

Let the length of the third side be $2x - 5$.

$$\begin{aligned} x + x + 45 + 2x - 5 &= 300 \\ 4x + 40 &= 300 \\ 4x &= 260 \\ x &= 65 \end{aligned}$$

$65 + 45 = 110$, $2 \times 65 - 5 = 125$.

$\therefore$ the sides are 65 m, 110 m and 125 m.

FOCUS on ...

1. The question: Asks you to find the dimensions of the triangular paddock.
2. The information: Gives you the cost of fencing and the relationship between the three sides of the paddock.
3. Your working: You need to find the perimeter of the paddock by dividing: $2340 \div 7.8 = 300$.
   You need to introduce a pronumeral to set up an equation to solve.
   Let the length of the first side = x.
   You add 45 to find the length of the second side: $x + 45$.
   The length of the third side is 5 less than twice x: $2x - 5$.
   The total of the side lengths equals the perimeter.
   $$\begin{aligned} x + x + 45 + 2x - 5 &= 300 \\ 4x + 40 &= 300 \\ 4x &= 260 \\ x &= 65 \end{aligned}$$
   $x + 45 = 65 + 45 = 110$, $2x - 5 = 2 \times 65 - 5 = 125$.
4. Your answer: Make sure you write the correct units:
   The lengths are 65 m, 110 m, 125 m.

A piece of copper wire a metre long is bent into the shape of a rectangle where the length of the rectangle is 10 cm longer than 4 times its width. What is the area of the rectangle?

Solution

Let the width of the rectangle be x

Let the length of the rectangle be $4x + 10$.

As 1 metre = 100 cm:

$$\begin{aligned} 2(x + 4x + 10) &= 100 \\ 2(5x + 10) &= 100 \\ 10x + 20 &= 100 \\ 10x &= 100 - 20 \\ 10x &= 80 \\ x &= 8 \end{aligned}$$

Also, $4 \times 8 + 10 = 42$.

$\therefore$ dimensions are 42 cm and 8 cm.

Area $= 42 \times 8 = 336$

$\therefore$ the area of the rectangle is 336 cm^2.

FOCUS on ...

1. The question: Asks you to find the area of the rectangle.
2. The information: Gives you the length of the wire and the relationship between the sides of the rectangle.
3. Your working: Change 1 metre to 100 cm.
   You need to introduce a pronumeral to set up an equation to solve.
   Let the width = x.
   The length is 4 times the width plus 10: $4x + 10$.
   The total of the side lengths equals the perimeter:
   $$\begin{aligned} 2(x + 4x + 10) &= 100 \\ 2(5x + 10) &= 100 \\ 10x + 20 &= 100 \\ 10x &= 100 - 20 \\ 10x &= 80 \\ x &= 8 \end{aligned}$$
   $4x + 10 = 4 \times 8 + 10 = 42$.
   The dimensions are 42 cm and 8 cm.
   Area $= 42 \times 8 = 336$
4. Your answer: Make sure you write the correct units:
   The area is 336 cm^2.

Now try these!

1. A triangular paddock is to be fenced at a cost of \$6.50 per metre. The longest side of the paddock is 25 metres longer than one side and 40 metres longer than the shortest side. Find the lengths of the three sides if the total fencing cost is \$1235.

2. A 60-cm length of wire is bent to form a rectangle where the length is 6 cm longer than twice the width. What is the area of the rectangle?

3. When Serena prints her photos she chooses a 1-cm white border to surround each image. The area of the white border is 20 cm^2. If the length of each photo's image is 4 cm longer than its width, what is the area of each photo print?

4. An existing fence is to be used as the fourth side (length) of a rectangular cattle yard. If the length is twice the width, what is the area of the yard, if 28 metres of fencing is required?

5. As fencing costs \$8.25/m, the cost of fencing a triangular paddock is \$2970. A side is 40 metres longer than one side and 70 metres shorter than the other side. Find the dimensions of the paddock.

6. CHALLENGE The cost of framing a rectangular picture is \$40 per metre. The length of the picture is 10 cm shorter than twice the width and the total cost of the framing is \$64. The picture is to be covered with glass which costs \$160 per metre^2. What is the cost of the glass?

Answers pages 134–135

KEY SKILL

16 Equations and geometry

HINTS

- Make sure you know the **Equations Hints** from Key Skill 13 on page 38.
- Complementary angles add to 90°, supplementary angles add to 180°.
- Angle sum of n-sided polygon is $180(n - 2)$ degrees.
- Each angle of regular n-sided polygon is $\frac{180(n-2)}{n}$ degrees.
- The sum of exterior angles of a convex polygon is 360°.

Examples

Gillian measures an angle where the difference between 3 times its complement and its supplement is 50°. What is the size of her angle?

Solution:

Let the angle be x.
The complement is $(90 - x)$.
The supplement is $(180 - x)$.

$$\begin{aligned} 3(90 - x) - (180 - x) &= 50 \\ 270 - 3x - 180 + x &= 50 \\ 90 - 2x &= 50 \\ 2x &= 90 - 50 \\ 2x &= 40 \\ x &= 20 \end{aligned}$$

∴ Gillian's angle is 20°. (Note: her angle could also be 70°.)

FOCUS on ...

1. The question: Asks you to find the size of Gillian's angle.
2. The information: Gives you the relationship between the angle's complement and supplement.
3. Your working: You need to introduce a pronumeral to set up an equation to solve.
   Let the angle = x.
   The complement is 90 − x, the supplement is 180 − x.
   $$\begin{aligned} 3(90 - x) - (180 - x) &= 50 \\ 270 - 3x - 180 + x &= 50 \\ 90 - 2x &= 50 \\ 2x &= 90 - 50 \\ 2x &= 40 \\ \frac{2x}{2} &= \frac{40}{2} \\ x &= 20 \end{aligned}$$
4. Your answer: Make sure you write the correct units: Gillian's angle is 20° or 70°.

In a triangle the largest angle is 3 times the size of a second angle, and the difference between the two smaller angles is 20°. Find the size of each angle.

Solution

Let the size of the second angle be x.
Let the size of the first angle be $3x$.
Let the size of the smallest angle be $x - 20$.
As angle sum of a triangle is 180°:

$$\begin{aligned} x + 3x + x - 20 &= 180 \\ 5x - 20 &= 180 \\ 5x &= 180 + 20 \\ 5x &= 200 \\ x &= 40 \end{aligned}$$

Other angles are $3 \times 40 = 120$, $40 - 20 = 20$.
∴ the angles are 120°, 40° and 20°.

FOCUS on ...

1. The question: Asks you to find the size of each angle.
2. The information: Gives you the relationship between the three angles.
3. Your working: You need to introduce a pronumeral to set up an equation to solve.
   Let the angle = x.
   The largest angle is 3 × x = 3x.
   The smallest angle is x − 20.
   The angle sum of a triangle is 180°:
   $$\begin{aligned} x + 3x + x - 20 &= 180 \\ 5x - 20 &= 180 \\ 5x &= 180 + 20 \\ 5x &= 200 \\ \frac{5x}{5} &= \frac{200}{5} \\ x &= 40 \end{aligned}$$
   Now, 3x = 3 × 40 = 120, and x − 20 = 40 − 20 = 20.
4. Your answer: Make sure you write the correct units: The angles are 120°, 40°, 20°.

Now try these!

1. Willow measured an angle where the difference between three times its complement and its supplement is 36°. What was the size of her angle?

2. In a triangle the largest angle is twice the smallest angle and the other angle is 20° larger than the smallest angle. What is the size of each of the three angles?

3. One angle of a parallelogram is 57° more than twice the size of another angle. What are the sizes of each angle in the parallelogram?

4. One angle of an isosceles triangle is 30° larger than another angle. What are the possible sizes of the angles of the triangle?

5. One angle in a right triangle is 10° more than 4 times the size of the smallest angle. What is the size of each of the angles?

6. CHALLENGE One angle of a triangle has its supplement 40° more than twice its complement. A second angle is 3 times the size of the third angle. What is the ratio of the angles of the triangle?

Answers page 135

KEY SKILL

17 Equations and speed

HINTS

- Make sure you know the **Equations Hints** from Key Skill 13 on page 38.
- Speed is a rate in which distance travelled is compared to the time taken, e.g. Travelling 80 km in 1 hour means 80 km/h.
- There are three formulae:
  Distance = Speed × Time
  Speed = Distance ÷ Time
  Time = Distance ÷ Speed

Examples

Daenerys left home at 8:30 am and travelled to Ballarat at an average speed of 80 km/h. Her husband Theon left home half an hour later and averaged a speed of 100 km/h. At what time did Theon catch up to Daenerys?

Solution

Let Daenerys's time of travel be x.

$\therefore$ Theon's time of travel is $x - \frac{1}{2}$

As Distance = Speed × Time:

$$80 \times x = 100(x - \tfrac{1}{2})$$
$$80x = 100x - 50$$
$$20x = 50$$
$$x = 2.5$$

8:30 am plus 2.5 h = 11 am.

$\therefore$ Theon caught up at 11 am.

FOCUS on ...

1. The question: Asks you to find the time when Theon caught up to Daenerys.
2. The information: Gives you the speed of both Daenerys and Theon, the time Daenerys left and how much later Theon started his trip.
3. Your working: You need to introduce a pronumeral to set up an equation to solve. Let the time of Daenerys's trip = x. As Theon left half an hour after Daenerys, his time = $(x - \frac{1}{2})$.
   Use the formula Distance = Speed × Time. The distance travelled by both drivers will be equal:
   $$\text{Distance} = 80 \times x = 100(x - \tfrac{1}{2})$$
   $$80x = 100x - 50$$
   $$20x = 50$$
   $$x = 2.5$$
   8:30 am plus 2.5 h = 11 am
4. Your answer: Make sure you write the correct units: Theon caught up at 11 am.

A plane leaves Adelaide airport at 15:50 and flies north-west at an average speed of 900 km/h. A second plane leaves the same airport at 16:35 and flies the same route. If the second plane averages 1000 km/h, how far will it fly before it overtakes the first plane?

Solution

Elapsed time = 16:35 − 15:50 = 45 min = 0.75 h

Let the first plane's time of flight be x.

$\therefore$ second plane's time of flight is $x - 0.75$

As Distance = Speed × Time:

$$900 \times x = 1000(x - 0.75)$$
$$900x = 1000x - 750$$
$$100x = 750$$
$$x = 7.5$$

After 7.5 hours, the second plane catches up.

Distance of first plane = 900 × 7.5 = 6750

$\therefore$ the planes have flown 6750 km.

FOCUS on ...

1. The question: Asks you to find the distance the planes have flown before the second catches up to the first.
2. The information: Gives you the speed of both planes and the times the two planes took off.
3. Your working: You need to find the time between take-offs: 16:35 minus 15:50 = 45 min, or 0.75 h. You need to introduce a pronumeral to set up an equation to solve. Let the time of the first plane's flight = x. As the second plane was 0.75 h later, its time = (x − 0.75).
   Use the formula Distance = Speed × Time.
   The distance travelled by both planes will be equal.
   $$\text{Distance} = 900 \times x = 1000(x - 0.75)$$
   $$900x = 1000x - 750$$
   $$100x = 750$$
   $$x = 7.5$$
   Use the formula again.
   Distance = 900 × 7.5 = 6750
4. Your answer: Make sure you write the correct units: The planes fly 6750 km.

Now try these!

1. Jayden left home at 10 am and travelled to Ipswich at an average speed of 60 km/h. His wife Mia left home an hour later and averaged 90 km/h. At what time did Mia catch up to Jayden?

2. Malia leaves a petrol station at 8:30 am and drives towards Mandura at an average speed of 70 km/h. Christian leaves the same petrol station at 8:50 am to try and catch Malia. If Christian drives at an average speed of 90 km/h, how far will he need to drive to catch up to Malia?

3. At 9 am a motorboat leaves a harbour and travels at an average speed of 12 km/h towards a small island. At 9:30 am a cabin cruiser leaves the same harbour and travels at an average speed of 20 km/h towards the same island. At what time will the cabin cruiser be alongside the motorboat?

4. Michael leaves Perth at 3 pm and drives towards Geraldton at an average speed of 80 km/h. Michelle leaves Geraldton at 4 pm and travels towards Perth at an average speed of 90 km/h. If Perth and Geraldton are 420 km apart, at what time, and where, will they meet?

5. Rhianna left for work at 8 am and drove at an average speed of 60 km/h. Unfortunately, she left her bag at home and so her flatmate got into his car 5 minutes after Rhianna left and drove after her at an average speed of 72 km/h. How long did it take for the flatmate to catch up to Rhianna?

6. CHALLENGE Mitch left home at 8:00 am and walked to school at an average speed of 5 km/h. When he arrived at school he realised that it was a pupil-free day, so he ran home at an average speed of 10 km/h. If he arrived home at 8:45 am, how far does he live from school and what was his average speed?

Answers pages 135–136

KEY SKILL

18 Equations and money

HINTS

- Make sure you know the **Equations Hints** from Key Skill 13 on page 38.
- The sum of the numbers a and $(b - a)$ is b, e.g. Two numbers add to 20. If one number is x, the other number is $20 - x$.

Examples

A school held an out-of-uniform day to raise funds for the Cancer Council. 900 students made a gold coin donation of \$1 or \$2 coins. If \$1420 was raised, how many more students donated a \$2 coin than a \$1 coin?

Solution

Let the number of \$1 coins be x.

$\therefore$ number of \$2 coins was $900 - x$.

$$1 \times x + 2(900 - x) = 1420$$
$$x + 1800 - 2x = 1420$$
$$1800 - x = 1420$$
$$x = 1800 - 1420$$
$$x = 380$$

Also, $900 - 380 = 520$

Difference $= 520 - 380 = 140$

$\therefore$ 140 more students gave a \$2 coin than those who gave a \$1 coin.

FOCUS on...

1. The question: Asks you to find how many more students gave a \$2 coin than a \$1 coin.
2. The information: Gives you the number of students who paid a gold coin and the total amount paid.
3. Your working: You need to introduce a pronumeral to set up an equation to solve. Let the number of \$1 coins = x. This means the quantity of \$2 coins = 900 − x. Adding the values of the \$1 and \$2 coins gives 1420.
$$1 \times x + 2(900 - x) = 1420$$
$$x + 1800 - 2x = 1420$$
$$1800 - x = 1420$$
$$x = 1800 - 1420$$
$$x = 380$$
$900 - 380 = 520$, then there were 380 \$1 and 520 \$2.
Difference $= 520 - 380 = 140$
4. Your answer: Make sure you write the correct units:
$\therefore$ 140 more students donated \$2, rather than \$1.

A shopkeeper sells bags of mixed nuts. She wants to mix cashews and almonds. Cashews cost \$24/kg and almonds \$18/kg. She plans to sell 120 kg of the mix. How many kilograms of each type of nut should be mixed for each mixed bag of nuts to cost the shopkeeper \$20/kg?

Solution

Let the mass of cashews be x.

$\therefore$ mass of almonds is $120 - x$

$$24 \times x + 18(120 - x) = 20 \times 120$$
$$24x + 2160 - 18x = 2400$$
$$6x = 240$$
$$x = 40$$

$\therefore$ 40 kg of cashews and 80 kg of almonds will be needed.

FOCUS on...

1. The question: Asks you to find the mass of each type of nut.
2. The information: Gives you the cost of each nut, the total mass of the nuts and the cost of each bag of nuts.
3. Your working: You need to introduce a pronumeral to set up an equation to solve. Let the mass (in kg) of cashews = x. This means the mass (in kg) of almonds = 120 − x. Multiply the cost of each nut by its mass and add. This will equal the total cost of all nuts.
Let the mass of cashews be x.
$\therefore$ mass of almonds is $120 - x$
$$24 \times x + 18(120 - x) = 20 \times 120$$
$$24x + 2160 - 18x = 2400$$
$$6x + 2160 = 2400$$
$$6x = 2400 - 2160$$
$$6x = 240$$
$$\frac{6x}{6} = \frac{240}{6}$$
$$x = 40$$
$120 - x = 120 - 40 = 80$
4. Your answer: Make sure you write the correct units:
$\therefore$ 40 kg of cashews and 80 kg of almonds will be needed.

Now try these!

1. A theatre sells children's tickets for $8 and adult tickets for $12. One evening 320 tickets were sold for a total of $3360. How many children and adults bought tickets?

2. Chocolate sultanas cost $12/kg and chocolate peanuts cost $16/kg. How many kilograms of each item should be used by a confectionery distributor if he wants to make a 20-kg mix which will cost him $15/kg?

3. For a pie-drive, Mary bought some apple pies and 3 times as many meat pies. If an apple pie costs $10 and a meat pie costs $8, how many meat pies did she buy if the total cost was $68?

4. A gourmet coffee shop has two different types of coffee. Batch A costs $24/kg and Batch B costs $32/kg. Anna purchased a 3-kg mixture of the batches and was charged $83.20. What mass of each batch was purchased?

5. At the start of his holidays Bruce spent $470 at the hardware store. He bought twice as many paintbrushes as rollers and 3 times as many cans of paint as paintbrushes. Each paintbrush cost $16, a roller cost $18 and a can of paint cost $70. How many paintbrushes did he buy?

6. CHALLENGE An amount of $6000 was split into two investment accounts attracting simple interest for a year. One amount earned 4% p.a. interest and the second amount earned 6% p.a. If the deposit in the 6% p.a. account earned $180 more interest than the deposit in the 4% p.a. account, how much was originally deposited in both accounts?

Answers page 136

KEY SKILL

19 Ratio and rates: Direct proportion

HINTS

- If one quantity (y) is directly proportional to another quantity (x), then $y \propto x$, or $y = kx$, where k is the constant of proportionality,

e.g. If $P = kn$, what is the value of k when $P = 12$ and $n = 4$? Subs. $P = 12$ and $n = 4$ in $P = kn$

$$12 = 4k$$
$$4k = 12$$
$$k = 3$$

Examples

The cost (C, in dollars) of carpet is directly proportional to the length (n, in metres). Evan purchased 16 metres of carpet for \$1792. Write an equation that links C and n and use it to find the cost of 21 metres of the same carpet.

Solution

Subs. $C = 1792$ and $n = 16$ in $C = kn$

$$1792 = 16k$$
$$16k = 1792$$
$$k = 112$$

$\therefore$ the equation is $C = 112n$

Subs $n = 21$ in $C = 112n$:

$$C = 112 \times 21 = 2352$$

$\therefore$ 21 metres of carpet costs \$2352.

FOCUS on...

1. The question: Asks you to find an equation that links C and n and use it to find the cost.
2. The information: Gives you the cost of a length of carpet and that C and n are directly proportional.
3. Your working: You need to substitute $C = 1792$ and $n = 16$ into $C = kn$ to find the constant k.
   $$1792 = k \times 16$$
   $$16k = 1792$$
   $$k = \frac{1792}{16}$$
   $$= 112$$
   The equation linking C and n is $C = 112n$.
   You need to substitute $n = 21$ in $C = 112n$.
   $$C = 112 \times 21$$
   $$= 2352$$
4. Your answer: Make sure you write the correct units: The carpet costs \$2352.

The height (h, in cm) of a certain termite mound is directly proportional to the square root of the number of termites (t). The height of a mound is 75 cm when the number of termites is 3600. Write an equation that links h and t and use it to find the number of termites in a mound that is 1.2 m high.

Solution

Subs. $h = 75$ and $t = 3600$ in $h = k\sqrt{t}$

$$75 = k \times \sqrt{3600}$$
$$60k = 75$$
$$k = 1.25$$

$\therefore$ the equation is $h = 1.25\sqrt{t}$

Subs. $h = 120$ in $h = 1.25\sqrt{t}$

$$120 = 1.25 \times \sqrt{t}$$
$$\sqrt{t} = \frac{120}{1.25}$$
$$\sqrt{t} = 96$$
$$t = 9216$$

$\therefore$ there are 9216 termites.

FOCUS on...

1. The question: Asks you to find an equation that links h and t and use it to find the number of termites.
2. The information: Gives you the height of a mound with a given number of termites and that h is directly proportional to the square root of t.
3. Your working: You need to substitute $h = 75$ and $t = 3600$ into $h = k\sqrt{t}$ to find the constant k.
   $$75 = k \times \sqrt{3600}$$
   $$60k = 75$$
   $$k = \frac{75}{60}$$
   $$= 1.25$$
   The equation linking h and t is $h = 1.25\sqrt{t}$.
   As 1.2 m is 120 cm, you need to substitute $h = 120$ in $h = 1.25\sqrt{t}$
   $$120 = 1.25\sqrt{t}$$
   $$\sqrt{t} = \frac{120}{1.25}$$
   $$= 96$$
   You square both sides:
   $$t = 96^2 = 9216$$
4. Your answer: Make sure you write the correct units: There are 9216 termites in the mound.

Now try these!

1 The amount (A, in \$) paid for a quantity of rice is directly proportional to the number (n) of bags purchased. Jerome bought 3 bags for \$35.70. Write an equation to link A and n, and use it to find the cost of 7 bags.

2 Molten metal is used to make a series of balls with varying diameters. The mass (M, in kg) of each ball varies directly with the cube of its diameter (d, in cm). A ball with diameter 2.4 cm has a mass of 6.912 kg. What is the mass of a ball with diameter 2.6 cm?

3 A spring stretches when a mass is attached to one end. The length of extension (E, in cm) is directly proportional to the magnitude of the mass (m, in g). When a mass of 40 g is attached, the extension is 0.8 cm. What mass will produce an extension of 2.5 cm?

4 The time (T, in seconds) taken for a simple pendulum to return to its starting position is directly proportional to the square root of its length (L, in mm). A pendulum of 2.5 metres takes 2 seconds. How long would it take a pendulum of 1.6 metres to return to its starting position?

5 The pressure P (in g/cm^2) exerted at any point on the base of a particular tank filled with water varies directly with the depth (d, in cm) of the liquid. If the tank is filled to a depth of 20 cm, the pressure exerted on the base of the tank is 8 g/cm^2. Find the pressure when another 10 cm of water is poured into the container.

6 CHALLENGE The stopping distance (D, in metres) of a skidding car is directly proportional to the square of the speed of the car (s, in km/h). At a speed of 16 km/h, a car will take 3.2 metres to stop. Sam is travelling at 100 km/h. If she braked and skidded, would she collide with a stationary car 120 metres ahead of her? Show your calculations.

Answers pages 136–137

REVISION TEST 5 Level of difficulty—Average

1. Stella is x years old now. Her brother Dominic is twice as old as Stella was 2 years ago. Write an expression for the sum of their ages in 10 years' time.

2. In a basketball game Shelley made 12 scoring shots consisting of three-point shots and two-point shots. If her total was 29, how many three-point shots did she make?

3. In a certain triangle the largest angle is twice the smallest angle and the other angle is 8° larger than the smallest angle. What is the size of each of the three angles?

4. Abigail left home at 7 am and travelled to Ballarat at an average speed of 80 km/h. Her husband Noah left home 15 minutes later and averaged 100 km/h. At what time did Noah catch up to Abigail?

5. Ben has $280 in cash in his wallet in $20 notes and $10 notes. There are 8 more $20 notes than $10 notes. How many $10 notes does Ben have in his wallet?

6. The length a spring extends is proportional to the mass of an object placed on the end of the spring. When a load of 3 kg is hung on the spring, the spring extends 8 cm. How far will the spring extend if a mass of 6 kg is attached?

Answers page 137

REVISION TEST  Level of difficulty—Challenging

1. Ava is travelling from her home to her aunt's home. For the first x km she averages 60 km/h. She then stops for a break for y minutes, before driving for another z km at an average speed of 90 km/h to arrive at her destination. How long does the trip take?

2. Lucas is paid $18 per hour for his part-time job. In one week he worked 14 hours which was a combination of normal time and double time. If he worked 2 hours more at double time than at normal time, how much was he paid for his work?

3. The cost of framing a rectangular picture is $50 per metre. The length of the picture is 8 cm shorter than twice the width and the total cost of the framing is $67. The picture is to be covered with glass which costs $140 per metre^2. What is the cost of the glass?

4. Candice walked from her apartment to the shops at an average speed of 6 km/h. When she arrived she realised that she had forgotten her wallet so she rode a friend's bike home to get it and then back to the shops at an average speed of 12 km/h. If her total travelling time was 40 minutes how far does she live from the shops?

5. Bethany was given $20 000 to invest. She invested some of the money in an account offering simple interest of 4.5% p.a. and the rest into another account offering simple interest of 4.25% p.a. At the end of 5 years she had earned total interest of $4425. How much had she originally invested in each account?

6. A man had a goose that laid golden eggs where the number of eggs laid was proportional to the square root of the number of hours the goose slept. When the goose slept for 9 hours, it laid 4 golden eggs. How long did the goose sleep when it laid 2 eggs?

Answers pages 137–138

KEY SKILL

20 Inequalities

HINTS

- An inequality is a mathematical sentence that uses >, < , ≥ or ≤.
- Inequalities (or inequations) are solved in the same way as equations except that the symbols are changed when multiplying or dividing by a negative number,

e.g. Solve $2a + 3 > 7$

$$2a > 7 - 3$$
$$2a > 4$$
$$a > 2$$

$$5 - 3x \geq 17$$
$$-3x \geq 17 - 5$$
$$\frac{-3x}{-3} \leq \frac{12}{-3}$$
$$x \leq -4$$

Reminder!

- Read the question carefully to identify what needs to be found.
- Re-read the question when you've finished your answer to make sure that the question has been answered and your solution makes sense.

Examples

Anna is 4 years younger than her brother Joshua. The sum of their ages is less than 20. By solving an inequality, what can be said about Joshua's age?

Solution

Let Joshua's age be x.

∴ Anna's age is $x - 4$.

$$x + x - 4 < 20$$
$$2x - 4 < 20$$
$$2x < 20 + 4$$
$$2x < 24$$
$$x < 12$$

∴ Joshua must be less than 12 years old.

FOCUS on ...

1. The question: Asks you to find a statement for Joshua's age.
2. The information: Gives you the relationship between Anna's age and Joshua's age, and a statement about the sum of their ages.
3. Your working: You need to introduce a pronumeral to set up an inequality to solve. Let Joshua's age = x. This means Anna's age = x − 4.
   $x + x - 4 < 20$
   $2x - 4 < 20$
   $2x < 20 + 4$
   $2x < 24$
   $x < 12$
4. Your answer: Make sure you write the correct units: Joshua must be less than 12 years old.

A drycleaner charges $7 for the first item and $5 for each extra item. What is the most number of items Kristin can have drycleaned for $40?

Solution

Let the number of extra items be x.

$$7 + 5x \leq 40$$
$$5x \leq 40 - 7$$
$$5x \leq 33$$
$$x \leq 6.6$$

∴ Kristin can take 6 extra items.

As 6 + 1 = 7, then she can take 7 items.

∴ Kristin can have 7 items drycleaned.

FOCUS on ...

1. The question: Asks you to find the maximum number of items Kristin can afford to get drycleaned.
2. The information: Gives you the price of drycleaning and the maximum amount Kristin will spend.
3. Your working: You need to introduce a pronumeral to set up an inequality to solve. Let the number of extra items after the first item = x.
   $7 + 5x \leq 40$
   $5x \leq 40 - 7$
   $5x \leq 33$
   $x \leq 6.6$
   ∴ Kristin can take 6 extra items.
   As 6 + 1 = 7, then she can take 7 items.
4. Your answer: Make sure you write the correct units: Kristin can dryclean 7 items.

Now try these!

1. Sophia is 3 years older than her sister Emma. The sum of their ages is less than 19. By solving an inequality, what can be said about Emma's age?

2. Angus has $200 to spend. He wants to buy as many shirts as possible and leave at least $90 for a pair of shoes. If the shirts cost $40 each, what is the most number of shirts he can possibly buy?

3. Jacob is 5 years older than William but 8 years younger than Ethan. If the sum of the three boys' ages is at most 99, find the possible maximum age of Jacob.

4. A shower-screen factory has a fixed operating cost of $2400 per week. If a shower screen costs $40 to produce and sells for $120, what is the minimum number of shower screens that should be sold each week to make a profit?

5. Alexis has $5400 in a savings account at the beginning of her trip. She has budgeted that she will spend $640 each week and that she will need at least $1300 for the plane trip home. How many weeks will Alexis's money last if she keeps to her budget?

6. CHALLENGE Sean builds boat trailers. He has costs totalling $6000 per month plus $2100 per trailer. He sells the trailers for $3200 each. What is the least number of trailers he must sell if he is to make a profit of at least $10 000 per month?

Answers page 138

KEY SKILL

21 Using formulae

HINTS

- A formula is a special type of equation that shows the relationship between different variables.
- In $A = 3p - q$, the independent variables are p and q and A is the dependent variable.
- The subject of a formula is the single variable (usually on the left-hand side), that everything else is equal to. 'To change the subject' means to rearrange the formula so that another variable is the subject of the formula.
- Formulae are used by substituting numbers for pronumerals, e.g. If $v = u + at$, find v if $u = 6$, $a = 10$, $t = 3$.

$$\begin{aligned} v &= 6 + 10 \times 3 \\ &= 36 \end{aligned}$$

Reminder!

- Read the question carefully to identify what needs to be found.
- Re-read the question when you've finished your answer to make sure that the question has been answered and your solution makes sense.

Examples

The formula $d = 4.9t^2$ is used to find the distance d, in metres, an object falls in t seconds, under the effect of gravity. How far does Dylan fall if he jumps out of a plane and waits 15 seconds before he opens his parachute?

Solution

Subs. $t = 15$ in $d = 4.9t^2$

$$\begin{aligned} &= 4.9 \times (15)^2 \\ &= 1102.5 \end{aligned}$$

$\therefore$ Dylan falls 1102.5 metres.

FOCUS on ...

1. The question: Asks you to find the distance Dylan falls.
2. The information: Gives you the formula linking distance (d) and time (t).
3. Your working: Subs. the value of t into the formula.

$$\begin{aligned} d &= 4.9t^2 \\ &= 4.9 \times (15)^2 \\ &= 1102.5 \end{aligned}$$

4. Your answer: Make sure you write the correct units: Dylan falls 1102.5 metres.

The volume V of a rectangular pyramid is found using the formula $V = \frac{lbh}{3}$, where l = length, b = breadth and h = height. Find the height of a rectangular pyramid with length 24 m, breadth 20 m and volume of 2160 m^3.

Solution

$$\begin{aligned} V &= \frac{lbh}{3} \\ 2160 &= \frac{24 \times 20 \times h}{3} \\ 2160 &= 160h \\ h &= \frac{2160}{160} \\ &= 13.5 \end{aligned}$$

$\therefore$ the height is 13.5 m.

FOCUS on ...

1. The question: Asks you to find the height.
2. The information: Gives you the formula linking volume with length, breadth and height. The volume (V), length (l) and breadth (b) are used to find the height (h).
3. Your working: Subs. the value of V, l and b into the formula.

$$\begin{aligned} V &= \frac{lbh}{3} \\ 2160 &= \frac{24 \times 20 \times h}{3} \\ 2160 &= 160h \\ h &= \frac{2160}{160} \\ &= 13.5 \end{aligned}$$

4. Your answer: Make sure you write the correct units: The height is 13.5 m.

Now try these!

1. Euler's formula $V + F = E + 2$ can be used to find the number of vertices (V), faces (F) and edges (E) on any polyhedron. How many faces has an icosohedron if it has 12 vertices and 30 edges?

2. A bucky-ball (or a buckminsterfullerene) is a molecule in the shape of a polyhedron made of hexagons and pentagons. It has 60 vertices and 90 edges. If it has 20 hexagonal faces, use Euler's theorem from question 1 to find the number of pentagonal faces.

3. If air resistance is ignored the distance (h, in metres) an object will fall to earth in t seconds is found using the formula $h = \frac{1}{2}gt^2$. Using gravity $g = 10$ m/s^2, how far will an object fall in 5 seconds?

4. Using $h = \frac{1}{2}gt^2$ from question 3, how long will it take an object to fall to earth from a height of 1 km? Give your answer correct to 3 significant figures.

5. The period (P, in seconds) is the time taken for a pendulum of length (L), in metres, to return to its starting point and can be found using the formula $P = 2\pi\sqrt{\frac{L}{g}}$.
Using gravity $g = 10$ m/s^2, find the period of a pendulum of length 6 metres, correct to one decimal place.

6. CHALLENGE Using $P = 2\pi\sqrt{\frac{L}{g}}$ from question 5, find the length of a pendulum if it takes 8 seconds to return to its starting point, to the nearest metre.

Answers pages 138–139

KEY SKILL

22 Linear relationships

HINTS

- A linear relationship between two variables occurs when one variable is directly proportional to another.
- When a linear relationship is graphed it appears as a straight line.
- The relationship is written as an equation and can be determined using a table or a number plane graph, e.g. Use the table to find the rule that links y in terms of x:

x	0	1	2	3	4
y	3	5	7	9	11

The rule is $y = 2x + 3$.

Examples

Pipe is sold according to its length and the following table details the cost ($\$C$) of different lengths ($n$, in metres).

n	3	5	8	20	100
C	7.05	11.75	18.8	47	235

Form an equation for C in terms of n and use the formula to find the length of pipe that can be purchased for \$37.60.

Solution

$$C = 2.35n$$

Subs. $C = 37.6$ in $C = 2.35n$

$$37.6 = 2.35n$$

$$n = \frac{37.6}{2.35} = 16$$

$\therefore$ 16 metres of pipe can be purchased.

FOCUS on ...

1. The question: Asks you to write the rule that links C and n, and use this equation to find the length (n) for a cost (C) of \$37.60.
2. The information: Gives you a series of values of n and the resultant values of C.
3. Your working: You need to use the table to find a rule linking C and n.
   The value of C is the value of n multiplied by 2.35.
   $C = 2.35n$
   Subs. $C = 37.6$ in $C = 2.35n$
   $37.6 = 2.35n$
   $n = \frac{37.6}{2.35}$
   $= 16$
4. Your answer: Make sure you write the correct units:
   16 metres of pipe is purchased.

A tap is dripping at a constant rate and the quantity of water in litres (W) is recorded over a period of time (t) in minutes.

t	30	60	90	120	150
W	6	12	18	24	30

Form an equation for W in terms of t and use the formula to find the amount of water, in kilolitres, wasted after one week.

Solution

$$W = \frac{t}{5}$$

$$\text{Mins in week} = 60 \times 24 \times 7$$
$$= 10\,080 \qquad \therefore 10\,080 \text{ minutes}$$

Subs. $t = 10\,080$ in $W = \frac{t}{5} = \frac{10\,080}{5}$

$\therefore$ 2016 litres, or 2.016 kL is wasted.

FOCUS on ...

1. The question: Asks you to write the rule that links W and t, and use this equation to find the water (W) wasted after one week.
2. The information: Gives you a series of values of t and the resultant value of W.
3. Your working: You need to use the table to find a rule linking W and t.
   The value of W is the value of t divided by 5:
   $W = \frac{t}{5}$
   Now, change time in one week to minutes.
   $60 \times 24 \times 7 = 10\,080$
   Subs. $t = 10\,080$ in $W = \frac{t}{5}$
   $= \frac{10\,080}{5} = 2016$
   Change to kilolitres by dividing by 1000.
   $2016 \div 1000 = 2.016$
4. Your answer: Make sure you write the correct units:
   2.016 kilolitres of water is wasted.

Now try these!

1 The cost (C), in dollars, of different lengths of metal (n), in metres, is listed in the table.

n	2	3	5	8	10	15
C	12	18	30	48	60	90

Form an equation for C in terms of n and use the formula to find the cost of 18 metres of material.

2 Brian uses the following table to advertise the charge (C), in dollars, of different periods of his lawnmowing (t), in hours.

t	1	1.5	2	2.5	3	5
C	50	60	70	80	90	130

Form an equation for C in terms of t and use the formula to find the cost of Brian's lawnmowing for 4 hours.

3 Inger buys 50 heart-shaped chocolates to place in gift baskets she is assembling for her market stall. The table shows the relationship between the number of gift baskets (n) and the number of chocolates (c) she has remaining.

n	0	2	5	7	10	12
c	50	44	35	29	20	14

Write an equation linking c and n, and find the number of chocolates she has remaining after making 16 baskets.

4 The cost (C), in dollars, of a hire car to travel different distances (d), in kilometres, is detailed in the table.

d	0	3	5	10
C	6.25	15.25	21.25	36.25

Form an equation for C in terms of d and use the formula to find the cost of hiring the car for a trip of 25 km.

5 Anatoly is travelling between two cities and the table shows the distance d, in kilometres, he still has to travel after driving for n hours. Anatoly is able to travel at 90 km/h for the entire journey.

n	1	2	4	6
d	530	440	260	80

Form an equation for d in terms of n and use the formula to find the distance he still has to travel after driving for 6 hours 15 minutes.

6 CHALLENGE At her local pool Sarah swam 300 metres on her first day, and on each following day increased the distance by 50 metres. Complete the table to show the relationship between days (n) and distance (D), in kilometres.

n	1	2	5	8
D				

Form an equation for D in terms of n and use the formula to find the distance, in kilometres, Sarah swam on the 25th day.

Answers page 139

KEY SKILL

23 Simultaneous equations A

HINTS

- An equation with two unknowns (e.g. $2x + y = 3$) has an infinite number of solutions. Two equations (e.g. $2x + y = 3$ and $x - 2y = -1$) can be solved simultaneously to find the value of x and y.
- There are three methods: algebraically by elimination or substitution, or graphically (using a number plane).

Examples

Two cousins compared their ages. The sum of their ages is 34 and the difference is 6. Find their ages.

Solution

Let the ages be x and y.

$$x + y = 34 \quad \ldots ①$$
$$x - y = 6 \quad \ldots ②$$

① + ②: $2x = 40$

$x = 20$

Subs. in ①: $20 + y = 34$

$y = 14$

$\therefore$ the cousins are 20 and 14.

FOCUS on ...

1. The question: Asks you to find the ages of the cousins.
2. The information: Gives you the relationship between the two ages.
3. Your working: Letting the two ages be x and y, set up a pair of simultaneous equations.
   Sum is 34: $x + y = 34 \ldots ①$
   Difference is 6: $x - y = 6 \ldots ②$
   Adding the two equations: $2x = 40$
   $x = 20$
   Substitute $x = 20$ in $x + y = 34$
   $20 + y = 34$
   $y = 14$
   $x = 20$ and $y = 14$
4. Your answer: Make sure you write the correct units:
   The cousins are 20 and 14.

The perimeter of a rectangle is 28 cm and the length is 5 cm more than half the width. Find the area of the rectangle.

Solution

Let the width be x cm and the length be y cm.

$$2(x + y) = 28$$
i.e. $x + y = 14 \quad \ldots ①$

Also, $y = \frac{x}{2} + 5 \quad \ldots ②$

Subs. ② in ①: $x + \frac{x}{2} + 5 = 14$

$x + \frac{x}{2} = 9$

$2x + x = 18$

$3x = 18$

$x = 6$

Subs. in ①: $6 + y = 14$

$y = 8$

$\therefore$ the dimensions are 8 cm and 6 cm.

$\therefore$ the area is 48 cm^2.

FOCUS on ...

1. The question: Asks you to find the area.
2. The information: Gives you the perimeter and the relationship between the length and the width.
3. Your working: Letting the width = x and length = y, the perimeter is $2(x + y) = 28$, or $x + y = 14 \ldots ①$
   Also, half of x is written as $\frac{x}{2}$, so $y = \frac{x}{2} + 5 \ldots ②$
   Substituting ② in ①: $x + \frac{x}{2} + 5 = 14$
   $x + \frac{x}{2} = 9$
   $2x + x = 18$
   $3x = 18$
   $x = 6$
   Substitute in ①: $6 + y = 14$
   $y = 14 - 6$
   $y = 8$
   Dimensions are 8 cm by 6 cm.
   Area $= 8 \times 6$
   $= 48$
4. Your answer: Make sure you write the correct units:
   The area is 48 cm^2.

Now try these!

1. Joachin and Leila are brother and sister. The sum of their ages is 22 and Joachin is 6 years older than his sister. Find the ages of the children.

2. The perimeter of a rectangular painting is 260 cm and the length is 20 cm less than twice the width. Find the area of the rectangle.

3. On a farm there are 40 animals. Some of the animals are chickens and the remainder are sheep. If there is a total of 128 feet, how many sheep are on the farm?

4. If twice the age of a daughter is added to her mother's age the sum is 50. When the age of the mother is subtracted from 5 times the daughter's age, the difference is 6. Find the ages of the mother and the daughter.

5. A group of 166 students attended the regional athletics carnival. The athletes travelled in 10 vehicles, some of them buses and the remainder cars. Each bus could carry 46 students and each car could hold 4 students. Find the number of buses and cars.

6. CHALLENGE OffSet Printers Company has two types of printers. Model A can print 1200 pamphlets per hour and Model B can print 1000 pamphlets per hour. Altogether OffSet Printers has 8 printers. If the firm can print 9000 pamphlets each hour, how many of each model is owned by the firm?

Answers page 139

KEY SKILL

24 Simultaneous equations B

HINTS

- Make sure you know the **Simultaneous equations Hints** from Key Skill 23 on page 60.

Examples

At the local show, James had 3 rides on the Crazy Coaster and 2 rides on the Extreme which cost a total of \$18.40. Grace spent 30 cents less than James when she had 2 rides on the Crazy Coaster and 3 rides on the Extreme. Find the cost of each ride.

Solution

Let the number of rides on the Crazy Coaster be c.
Let the number of rides on the Extreme be e.

$3c + 2e = 18.4$ … ①
$2c + 3e = 18.1$ … ②
2 × ① : $6c + 4e = 36.8$ … ③
3 × ② : $6c + 9e = 54.3$ … ④
④ − ③ : $5e = 17.5$
$e = 3.5$
Subs. in ① : $3c + 2(3.5) = 18.4$
$3c + 7 = 18.4$
$3c = 11.4$
$c = 3.8$

∴ Crazy Coaster costs \$3.80, Extreme costs \$3.50.

FOCUS on ...

1. The question: Asks you to find the cost of each ride.
2. The information: Gives you the rides and the costs for two people.
3. Your working: Letting the cost of the Crazy Coaster ride = c and the cost of the Extreme ride = e, set up a pair of simultaneous equations.
   $3c + 2e = 18.4$ … ①
   $2c + 3e = 18.1$ … ②
   By multiplying the equations you can eliminate c:
   2 × ① : $6c + 4e = 36.8$ … ③
   3 × ② : $6c + 9e = 54.3$ … ④
   ④ − ③ : $5e = 17.5$
   $e = 3.5$
   Subs. in ① : $3c + 2(3.5) = 18.4$
   $3c + 7 = 18.4$
   $3c = 18.4 - 7$
   $3c = 11.4$
   $c = 3.8$
4. Your answer: Make sure you write the correct units: The Crazy Coaster ride costs \$3.80 and the Extreme ride costs \$3.50.

Each morning that he exercises, Luke cycles at the same speed or runs at the same speed. On Monday he cycled for half an hour and ran for an hour and covered 18 km. On Wednesday he cycled for three-quarters of an hour and ran for half an hour and covered a distance of 19 km. What speed does he run and cycle?

Solution

Let Luke's cycling speed be c km/h.
Let Luke's running speed be r km/h.

Monday: $\frac{c}{2} + r = 18$ … ①
Wednesday: $\frac{3c}{4} + \frac{r}{2} = 19$ … ②
2 × ② : $\frac{3c}{2} + r = 38$ … ③
③ − ① : $c = 20$
Subs. in ① : $r = 8$

∴ Luke cycles at 20 km/h and runs at 8 km/h.

FOCUS on ...

1. The question: Asks you to find Luke's cycling speed and his running speed.
2. The information: Gives you two days' times and distances.
3. Your working: You use Distance = Speed × Time to write two simultaneous equations for Luke's distances. Also, let the cycling speed = x and the running speed = y.
   $\frac{c}{2} + r = 18$ … ①
   $\frac{3c}{4} + \frac{r}{2} = 19$ … ②
   By multiplying the equation ② you can eliminate r.
   2 × ② : $\frac{3c}{2} + r = 38$ … ③
   ③ − ① : $\frac{3c}{2} - \frac{c}{2} = 20$
   $c = 20$
   Subs. in ① : $\frac{20}{2} + r = 18$
   $r = 18 - 10 = 8$
4. Your answer: Make sure you write the correct units: Luke cycles at 20 km/h and runs at 8 km/h.

Now try these!

1. Three milkshakes and 2 juices cost $24.40 while 2 milkshakes and 3 juices cost $22.60. What is the cost of each drink?

2. Each weekend that she exercises Jodie jogs at a consistent speed or walks at a consistent speed. On Saturday she jogged for half an hour and then walked for 20 minutes and covered 8 km. On the following day she jogged for 20 minutes and walked for half an hour and covered 7 km. At what speed does Jodie walk and jog?

3. A money jar contains 60 coins with a total value of $6.40. There are twice as many 10-cent coins as 5-cent coins and the remaining coins are 20-cent pieces. How many 20-cent coins are in the jar?

4. Two runners start from the same point at the same time. They will be 8 km apart after 2 hours if they run in the same direction, or they will be 20 km apart after one hour if they run in opposite directions. What is the speed of both runners?

5. A concert had the following ticket prices: Adults $20, Pensioners $14, Children $10. Three times as many children as pensioners attended the concert. If 360 tickets were sold and ticket sales totalled $5760, how many adult tickets were sold for the concert?

6. CHALLENGE Each afternoon Enrico rides the 20 km from work to home but one day he got a flat tyre and had to walk the rest of the distance. He rides at an average speed of 30 km/h but walks at 6 km/h. If the entire journey took one hour, how far did he have to walk?

Answers pages 139–140

KEY SKILL

25 Simultaneous equations C

HINTS

- When solving simultaneous equations graphically, the solution is the point of intersection.
- Make sure you know **the Linear relationships** Hints from Key Skill 22 on page 58.
- Make sure you know the **Simultaneous equations Hints** from Key Skills 23 and 24 on pages 60 and 62.

Examples

The sum of two boys' ages is 11 and the difference is 3. Write two equations and solve them graphically to find the ages of the boys.

Solution

Let the two ages be x and y.

$x + y = 11$

$\therefore y = 11 - x$

x	0	1	2
y	11	10	9

$x - y = 3$

$\therefore y = x - 3$

x	3	4	5
y	0	1	2

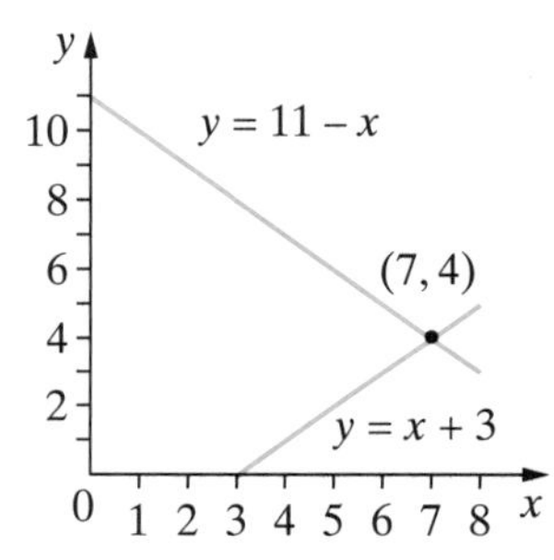

$\therefore$ the ages of the two boys are 7 and 4.

FOCUS on ...

1. The question: Asks you to write two equations and solve them simultaneously.
2. The information: Gives you the sum and difference of the two ages.
3. Your working: Letting the two ages be x and y, set up a pair of simultaneous equations.
   Sum: x + y = 11, or y = 11 − x
   Difference: x − y = 3, or y = x − 3
   For each equation, complete a table of values and graph the two lines on the same number plane (see opposite).
   The point of intersection of the two lines gives you the solution: the point (7, 4) means the boys are 7 and 4.
4. Your answer: Make sure you write the correct units: The boys are 7 and 4.

Two taxis charge different rates. Taxi A charges a base rate of \$10 plus \$2 per kilometre, while Taxi B charges \$4 per kilometre. By drawing two graphs, find the length of a taxi ride that will cost the same in both taxis.

Solution

Let the number of kilometres be x.

Taxi A: $C = 10 + 2x$

x	1	2	3
C	12	14	16

Taxi B: $C = 4x$

x	1	2	3
C	4	8	12

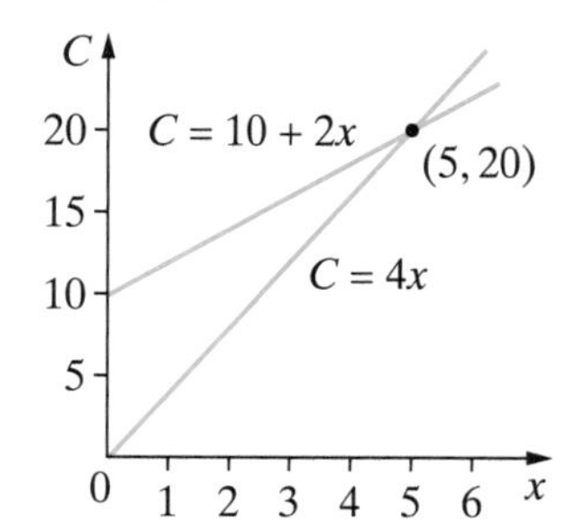

$\therefore$ a 5-km ride in each taxi costs the same amount.

FOCUS on ...

1. The question: Asks you to write two equations and solve them simultaneously.
2. The information: Gives you the costs of two taxis.
3. Your working: Let the cost of the taxi fare be C and the number of kilometres be x and set up a pair of simultaneous equations for each taxi.
   Taxi A: C = 10 + 2x
   Taxi B: C = 4x
   For each equation, complete a table of values and graph the two lines on the same number plane (see opposite).
   The point of intersection of the two lines gives you the solution: the point (5, 20) means the distance of the fare is 5 km (for a cost of \$20).
4. Your answer: Make sure you write the correct units: The distance is 5 km.

Now try these!

1. Gemma is older than her sister Phoebe. The sum of their ages is 12 and the difference is 4. Write two equations and solve them graphically to find the ages of the girls.

2. Two electricians charge different rates. Barry charges a call-out fee of $60 plus $40 per hour. Garry charges a flat rate of $60 per hour. By the use of graphs, find the length of a job (in hours) if Barry and Garry charge the same amount.

3. Simone and Kate are saving for their Bali holiday. Simone has $600 and is saving at $100 per week, while Kate has $500 and is saving at $120 per week. Write two equations representing the amount saved by each girl and solve the equations graphically to find the time it will take until they have saved the same amount.

4. Two adult tickets and a child's ticket to a concert cost a total of $20, while an adult ticket and four children's tickets cost a total of $24. By drawing two graphs, find the cost of each ticket.

5. Bucket A has 6 litres of water but is leaking at the rate of 400 mL/minute. The water is leaking into Bucket B which contains 4 litres of water. Draw tables to graph the amounts of water in each bucket and determine when the buckets have the same amount of water.

6. CHALLENGE Ethan and Jake are going to race their bikes. Jake leaves at 9:30 am, gets a 45-minute head start and rides at a constant speed of 12 km/h. When Ethan starts, he rides at a constant speed of 16 km/h. Draw graphs to represent the distance travelled and find the time when Ethan overtakes Jake.

Answers pages 140–141

KEY SKILL

26 Quadratic equations A

HINTS

- Expressions of the form $ax^2 + bx + c$ are called quadratic trinomials.
- Sometimes a quadratic can be factorised into the form $(x - p)(x - q)$.
- If $(x - p)(x - q) = 0$, then $x = p$ or $x = q$.
  e.g. Solve $x^2 - 3x - 4 = 0$
  $(x - 4)(x + 1) = 0$
  $x = 4$ or -1

Examples

The product of two consecutive even numbers is 48. Form an equation and solve it to find the numbers.

Solution

Let the two numbers be x and $x + 2$.

$$\begin{aligned} x(x + 2) &= 48 \\ x^2 + 2x &= 48 \\ x^2 + 2x - 48 &= 0 \\ (x + 8)(x - 6) &= 0 \\ x &= -8 \text{ or } 6 \end{aligned}$$

As $x = -8$, then $x + 2 = -6$.
As $x = 6$, then $x + 2 = 8$.
$\therefore$ the numbers are -8 and -6 or 6 and 8.

FOCUS on ...

1. The question: Asks you to find the consecutive numbers.
2. The information: Gives you the value of the product of the two consecutive numbers.
3. Your working: You need to introduce a pronumeral to set up an equation to solve.
   Let the first number = x.
   Add 2 to x to write the next even number: x + 2.
   You multiply to find the product.
   $$\begin{aligned} x(x + 2) &= 48 \\ x^2 + 2x &= 48 \\ x^2 + 2x - 48 &= 0 \\ (x + 8)(x - 6) &= 0 \\ x &= -8 \text{ or } 6 \end{aligned}$$
   As $x = -8$, then $x + 2 = -6$.
   As $x = 6$, then $x + 2 = 8$.
   The numbers are -8 and -6 or 6 and 8.
4. Your answer: The numbers are -8 and -6 or 6 and 8.

The length of a rectangle is 6 cm more than its width. If the area is 27 cm², what is the perimeter of the rectangle?

Solution

Let the width of the rectangle be x.
Let the length of the rectangle be $x + 6$.

$$\begin{aligned} x(x + 6) &= 27 \\ x^2 + 6x - 27 &= 0 \\ (x + 9)(x - 3) &= 0 \\ x &= -9 \text{ or } 3 \text{ (but } x \neq -9) \end{aligned}$$

As $x = 3$, then $x + 6 = 9$.
$\therefore$ dimensions are 9 cm by 3 cm.

$$\begin{aligned} \text{Perimeter} &= 2(9 + 3) \\ &= 24 \end{aligned}$$

$\therefore$ the perimeter is 24 cm.

FOCUS on ...

1. The question: Asks you to find the perimeter.
2. The information: Gives you the area and the relationship between the two sides.
3. Your working: You need to introduce a pronumeral to set up an equation to solve.
   Let the width = x.
   Add 6 to x to find the length: x + 6.
   You multiply to find the area:
   $$\begin{aligned} x(x + 6) &= 27 \\ x^2 + 6x - 27 &= 0 \\ (x + 9)(x - 3) &= 0 \\ x &= -9 \text{ or } 3 \end{aligned}$$
   But as x = width, then $x > 0$, so $x \neq -9$.
   As $x = 3$, then $x + 6 = 9$.
   The dimensions are 9 cm and 3 cm.
   $$\begin{aligned} \text{Perimeter} &= 2(9 + 3) \\ &= 2 \times 12 \\ &= 24 \end{aligned}$$
4. Your answer: Make sure you write the correct units:
   The perimeter is 24 cm.

Now try these!

1. The product of two consecutive odd numbers is 143. Form an equation and solve it to find the two numbers.

2. The length of a rectangle is 5 cm longer than its width. If the area of the rectangle is 84 cm^2 find the perimeter.

3. The sum of two numbers is 14 and the sum of their squares is 100. What are the two numbers?

4. A rectangular paddock has an area of 2400 m^2. The width is 2 metres shorter than its length. If the paddock is to be fenced at $2.90 per metre, what will be the cost of the fence?

5. A rectangular garden of length 14 metres by 8 metres is surrounded by a grass path of consistent width. If the total area of the garden and path is 216 m^2, what is the width of the grass path?

6. CHALLENGE The square root of 1 less than a certain number is the same as the number subtracted from 7. What is the number?

Answers page 141

KEY SKILL

27 Quadratic equations B

HINTS

- Make sure you know the **Quadratic equations Hints** from Key Skill 26 on page 66.
- A **non-monic quadratic** is of the form $ax^2 + bx + c$, where $a \neq 1$ or 0

e.g. Solve $3x^2 - 5x - 2 = 0$

$$(3x + 1)(x - 2) = 0$$
$$x = -\frac{1}{3}, 2$$

Examples

The base of a parallelogram is 3 cm longer than twice its perpendicular height. If the area of the parallelogram is 65 cm², what is the length of its base?

Solution

Let the perpendicular height of the parallelogram be x.

$\therefore$ the length of the parallelogram is $2x + 3$.

$$x(2x + 3) = 65$$
$$2x^2 + 3x = 65$$
$$2x^2 + 3x - 65 = 0$$
$$(2x + 13)(x - 5) = 0$$
$$x = -6.5 \text{ or } x = 5$$
$$(\text{but } x > 0)$$

As $x = 5$, then $2x + 3 = 13$.

$\therefore$ the base is 13 cm.

FOCUS on...

1. The question: Asks you to find the length of the base.
2. The information: Gives you the area and the relationship between the base and perpendicular height.
3. Your working: You need to introduce a pronumeral to set up an equation to solve.
   Let the perpendicular height = x.
   You double x and add 3 to find the base: 2x + 3.
   As A = bh, you multiply to find the area:
   $$x(2x + 3) = 65$$
   $$2x^2 + 3x = 65$$
   $$2x^2 + 3x - 65 = 0$$
   $$(2x + 13)(x - 5) = 0$$
   $$x = -6.5, 5$$
   But as x = height, then $x > 0$, so $x \neq -6.5$.
   As x = 5, then $2x + 3 = 2 \times 5 + 3 = 13$.
4. Your answer: Make sure you write the correct units:
   The base of the parallelogram is 13 cm.

The hypotenuse of a triangle is 4 cm longer than three times its shortest side. If the third side is 24 cm, what is the area of the triangle? (See Key Skill 28.)

Solution

Let the shortest side be x.

$\therefore$ the hypotenuse is $3x + 4$.

Using Pythagoras' theorem:

$$(3x + 4)^2 = x^2 + 24^2$$
$$9x^2 + 24x + 16 = x^2 + 576$$
$$8x^2 + 24x - 560 = 0$$
$$x^2 + 3x - 70 = 0$$
$$(x + 10)(x - 7) = 0$$
$$x = -10 \text{ or } 7$$
$$(\text{but } x > 0)$$

As $x = 7$, then $3x + 4 = 25$.

The sides are 7 cm, 24 cm, 25 cm.

$$\text{Area} = \frac{1}{2} \times 7 \times 24 = 84$$

$\therefore$ area is 84 cm².

FOCUS on...

1. The question: Asks you to find the area of a triangle.
2. The information: Gives you the length of one side and the relationship between the other two sides.
3. Your working: You need to introduce a pronumeral to set up an equation to solve.
   Let the shortest side = x.
   You multiply x by 3 and add 4 to find the length of the hypotenuse: 3x + 4.
   The three sides are x, 24, 3x + 4, and using Pythagoras:
   $$(3x + 4)^2 = x^2 + 24^2$$
   $$(3x + 4)(3x + 4) = x^2 + 24^2$$
   $$9x^2 + 24x + 16 = x^2 + 576$$
   $$8x^2 + 24x - 560 = 0$$
   $$x^2 + 3x - 70 = 0$$
   $$(x + 10)(x - 7) = 0$$
   $$x = -10 \text{ or } 7$$
   But as x = side length, then $x > 0$, so $x \neq -10$.
   As x = 7, then $3x + 4 = 3 \times 7 + 5 = 25$.
   Shorter sides are 7 cm and 24 cm (hypotenuse = 25 cm).
   $\text{Area} = \frac{1}{2} \times 7 \times 24 = 84$
4. Your answer: Make sure you write the correct units:
   The area is 84 cm².

Now try these!

1. The base of a triangle is 4 cm longer than twice the perpendicular height. If the area of the triangle is 120 cm^2, what is the length of the base?

2. One side of a right triangle is 2 cm longer than the shortest side and 2 cm shorter than the longest side. What is the area of the triangle?

3. An equilateral triangle and a rectangle have the same perimeter. The area of the rectangle is 24 cm^2 and its width is 1 cm shorter than half its length. What is the length of each side of the triangle?

4. The dimensions of a square are modified to make a rectangle. Two opposite sides of a square are increased by 6 cm and the other two sides are decreased by 4 cm. The area of the rectangle is the same as the area of the original square. What was the perimeter of the square?

5. Two squares are drawn. The side of the larger square is 2 cm longer than the side of the smaller square. If the sum of the areas of the squares is 100 cm^2, what is the difference in their areas?

6. CHALLENGE An open box is to be formed out of a rectangular piece of cardboard whose length is 8 cm longer than its width. To form the box, squares of side 2 cm are removed from the four corners of the cardboard. If the volume of the box is 168 cm^3, what was the area of the original piece of cardboard?

Answers pages 141–142

REVISION TEST  Level of difficulty—Average

1. All rides at an amusement park cost $3. Donald has $32 to spend but needs to buy a big teddy bear for $7 for his sister. Solve an inequality to find the maximum number of rides he can buy.

2. The surface area (S) of a hemisphere with given radius (r) is found using the formula $S = 3\pi r^2$. Find the exact radius of a hemisphere, with a surface area of $\frac{64\pi}{3}\text{cm}^2$.

3. Each time Christine cleans her car she uses the same amount of car shampoo. The table shows the relationship between the amount (A, in mL) remaining in the bottle and the number of times (n) she has washed her car.

n	0	2	4	6	8
A	400	360	320	280	240

Write an equation linking A and n, and find the amount remaining after 16 car washes.

4. Pablo loves to go to the movies. Last vacation he watched 10 films and spent $159. If a ticket to a 3D movie cost $18 and a 2D movie ticket cost $15, how many 3D movies did Pablo see?

5. Avery and Summer jog each morning. Avery ran 2 km this morning and plans to increase the distance each week by 500 metres. Summer ran 1.2 km this morning and plans to increase her distance each week by 700 metres. Write equations for each girl's distance and solve the equations graphically to find the number of weeks until both girls run the same distance.

6. One side of a rectangular paddock is 2 metres longer than the other side. If the area of the paddock is 360 m^2, what are the dimensions of the paddock?

Answers page 142

REVISION TEST  Level of difficulty—Challenging

1. Sean sells second-hand lawnmowers. He has costs totalling \$420 per month plus \$80 per lawnmower. He sells the lawnmowers for \$210 each. What is the least number of lawnmowers he must sell if he is to make a profit of at least \$2000 per month?

2. Last week Fiona bought a dress and a pair of jeans for \$260. This week the dress is discounted by 10% and the jeans by 15%. If she had bought the items this week she would have saved a total of \$30. How much did she pay for each item?

3. Given $t = \dfrac{\sqrt{u^2 + 2as} - u}{a}$, find the value of s when $t = 1.5$, $u = 18$, $a = 9.8$.

4. One diagonal of a rhombus is 4 cm longer than the other diagonal. If the area of the rhombus is 96 cm^2, find the perimeter.

5. The numerator of a fraction is 4 more than the denominator. If the value of the fraction is 2 less than the value of the denominator, what are the two possible values of the denominator?

6. A sand-timer has been made where sand from a top container falls into a lower container. At 7:20 am, the top container has 960 cm^3 of sand and is losing sand at the rate of 15 cm^3 per minute. The lower container has 600 cm^3 of sand. Draw tables to graph the amounts of sand in each container and determine at what time the containers have the same amount of sand.

Answers page 143

KEY SKILL

MEASUREMENT AND GEOMETRY

28 Pythagoras' theorem A

HINTS

- In a right-angled triangle the square of the hypotenuse equals the sum of the squares of the other two sides:

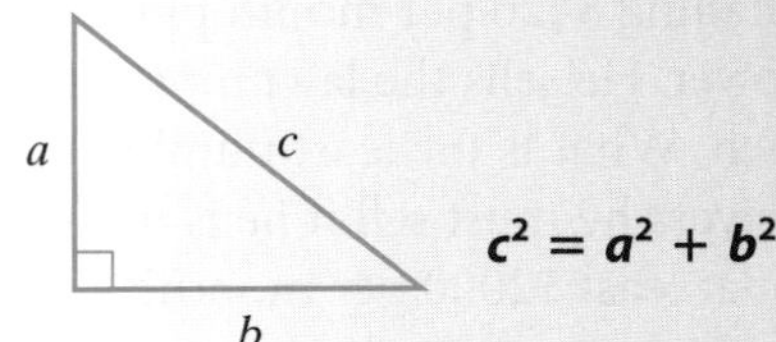

$c^2 = a^2 + b^2$

Examples

A square has an area of 16 cm². A circle is drawn through the vertices of the square. Find the circumference of the circle, correct to 2 decimal places.

Solution

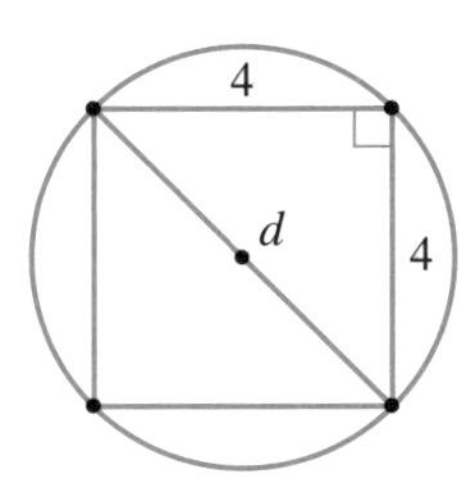

$$\begin{aligned} \text{Area} &= 16 \text{ cm}^2 \\ \text{Side} &= \sqrt{16} \\ &= 4 \\ d^2 &= 4^2 + 4^2 \\ &= 32 \\ d &= \sqrt{32} \\ \text{Circumference} &= \pi d \\ &= \pi \times \sqrt{32} \\ &= 17.771\,531\,75 \ldots \\ &= 17.77 \text{ (2 dec. pl.)} \end{aligned}$$

∴ the circumference is 17.77 cm.

FOCUS on ...

1. The question: Asks you to find the circumference.
2. The information: Gives you the area of a square.
3. Your working: Draw a diagram showing a square inside a circle.
   As the area of the square is 16 cm², then the side length is 4 cm.
   Let the diameter = d, and use Pythagoras' theorem:
   $$\begin{aligned} d^2 &= 4^2 + 4^2 \\ &= 32 \\ d &= \sqrt{32} \end{aligned}$$
   To find circumference, use $C = \pi d$
   $$\begin{aligned} &= \pi \times \sqrt{32} \\ &= 17.771\,531\,75 \ldots \\ &= 17.77 \text{ (2 dec. pl.)} \end{aligned}$$
4. Your answer: Make sure you write the correct units:
   The circumference is 17.77 cm.

Nadine has a paddock in the shape of an equilateral triangle with perimeter of 1.05 km. What is the area of the paddock, to the nearest hectare?

Solution

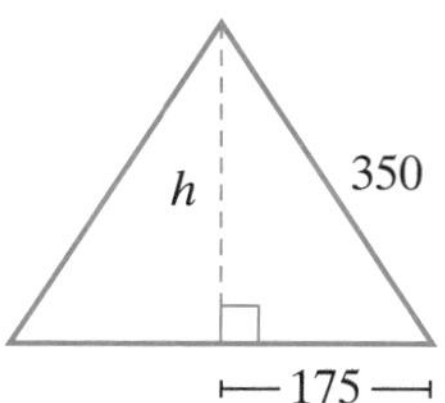

$1.05 \div 3 = 0.35$

∴ sides are 350 m.

$$\begin{aligned} h^2 &= 350^2 - 175^2 \\ &= 91\,875 \\ h &= \sqrt{91\,875} \\ \text{Area} &= \frac{1}{2} \times 350 \times \sqrt{91\,875} \\ &= 53\,044 \text{ (nearest whole)} \end{aligned}$$

$53\,044 \div 10\,000 = 5$ (nearest whole)

∴ the area of the paddock is 5 hectares.

FOCUS on ...

1. The question: Asks you to find the area.
2. The information: Gives you the perimeter of an equilateral triangle.
3. Your working: Draw a diagram showing an equilateral triangle with a perpendicular height.
   As the perimeter is 1.05 km, then you divide by 3 to find the length of each side: $1.05 \div 3 = 0.35$.
   Rewrite as 350 m. The perpendicular bisects the side, so you have a right triangle. Let height = h and use Pythagoras' theorem:
   $$\begin{aligned} h^2 &= 350^2 - 175^2 \\ &= 91\,875 \\ h &= \sqrt{91\,875} \\ \text{Area} &= \frac{1}{2} \times 350 \times \sqrt{91\,875} \\ &= 53\,044 \text{ (nearest whole)} \end{aligned}$$
   The area is 53 044 m².
   You divide by 10 000 to change to hectares:
   $53\,044 \div 10\,000 = 5.3044 = 5$ (nearest whole).
4. Your answer: Make sure you write the correct units:
   The area is 5 hectares.

Now try these!

1. A square has an area of 36 cm^2. A circle is drawn through the vertices of the square. What is the area of the circle, in terms of π?

2. An equilateral triangle has a perimeter of 48 cm. Find the area of the triangle, correct to 3 significant figures.

3. When Lachlan drives to work he picks up his workmate Ryan. Lachlan lives directly 18 km south of Ryan and their workplace is 15 km directly east of Ryan's house. Lachlan always drives at 40 km/h. How much time, to nearest minute, would Lachlan save if he did not have to pick up Ryan? (Assume there is a direct road from his home to work.)

4.

50

60

80

The diagram shows a paddock. If fencing costs \$7.65/metre, find the cost of fencing the paddock, to the nearest dollar.

5. A garden is in the shape of a rhombus whose side is 30 metres and one diagonal is 48 metres. Find the area of the garden.

6. CHALLENGE A sign is attached to the outside of a building. From a point on the ground 20 metres from the base of a building, the distance to the top of a sign is 28 metres, and the distance to the bottom of the sign is 25 metres. If the sign is 10 metres wide, what is the area of the sign, to the nearest square metre?

Answers pages 143–144

KEY SKILL

29 Pythagoras' theorem B

HINTS

- Make sure you know the **Pythagoras' theorem A Hints** from Key Skill 28 on page 72.

Examples

Two ships leave port at the same time. Ship A is heading west and Ship B is heading south. After 8 hours the ships are 400 km apart. If Ship A travels at an average speed of 40 km/h, what is the average speed of Ship B?

Solution

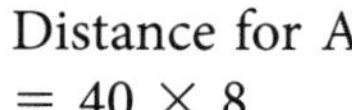

Distance for A
$= 40 \times 8$
$= 320$

$x^2 = 400^2 - 320^2$
$= 57\,600$
$x = 240$
$\therefore$ Ship B travelled 240 km
Speed for B $= 240 \div 8$
$= 30$
$\therefore$ Ship B averaged 30 km/h.

FOCUS on ...

1. The question: Asks you to find the speed of Ship B.
2. The information: Gives you the direction of the ships, the distance apart after a period of time and the speed of Ship A.
3. Your working: Draw a diagram to show the two ships.
   Using D = ST, you multiply to find the distance travelled by Ship A: $40 \times 8 = 320$.
   Apply Pythagoras' theorem:
   $x^2 = 400^2 - 320^2$
   $= 57\,600$
   $x = \sqrt{57\,600}$
   $= 240$
   Ship B had travelled 240 km.
   Using $S = \frac{D}{T}$, you divide to find the speed of Ship B.
   $240 \div 8 = 30$
4. Your answer: Make sure you write the correct units: Ship B averaged 30 km/h.

The sides of a right triangle are x, $(2x + 2)$ and $(2x + 3)$. Find the length of the shortest side, and hence the area of the triangle.

Solution

$x^2 + (2x + 2)^2 = (2x + 3)^2$
$x^2 + 4x^2 + 8x + 4 = 4x^2 + 12x + 9$
$5x^2 + 8x + 4 = 4x^2 + 12x + 9$
$x^2 - 4x - 5 = 0$
$(x - 5)(x + 1) = 0$
$x = 5, -1$ (but $x > 0$)
$\therefore x = 5$
Now, $2x + 2 = 12$
$2x + 3 = 13$
Dimensions are 5, 12, 13.
Area $= \frac{1}{2} \times 5 \times 12$
$= 30$
$\therefore$ area is 30 units^2.
$\therefore$ shortest side is 5 units, area is 30 units^2.

FOCUS on ...

1. The question: Asks you to find the length of the shortest side and the area.
2. The information: Gives you the three sides of a right triangle in terms of x.
3. Your working: The shortest side is x and the longest side (hypotenuse) is (2x + 3).
   Draw a diagram to show the three sides.
   Apply Pythagoras' theorem:
   $x^2 + (2x + 2)^2 = (2x + 3)^2$
   $x^2 + 4x^2 + 8x + 4 = 4x^2 + 12x + 9$
   $5x^2 + 8x + 4 = 4x^2 + 12x + 9$
   $x^2 - 4x - 5 = 0$
   $(x - 5)(x + 1) = 0$
   $x = 5, -1$ (but $x > 0$)
   $\therefore x = 5$
   $2x + 2 = 2 \times 5 + 2 = 12$
   $2x + 3 = 2 \times 5 + 3 = 13$
   The dimensions are 5 units, 12 units, 13 units.
   Area $= \frac{1}{2} \times 5 \times 12 = 30$
4. Your answer: Make sure you write the correct units: The shortest side is 5 units and area is 30 units^2.

Now try these!

1 Two boats leave a harbour at the same time. Boat X heads north-east at an average speed of 15 km/h while Boat Y travels south-east at 20 km/h. How far apart are the two boats after 4 hours?

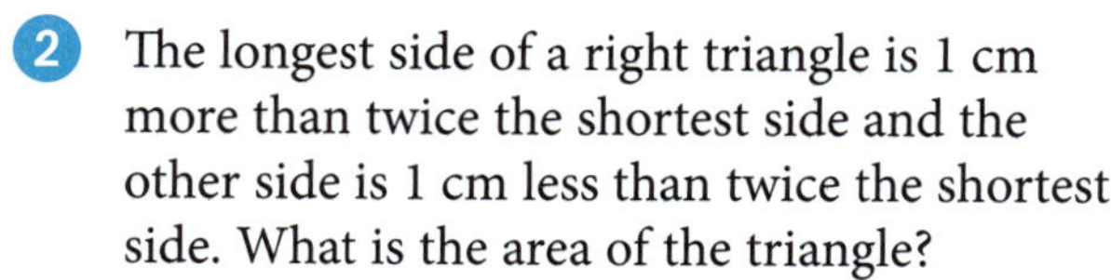

2 The longest side of a right triangle is 1 cm more than twice the shortest side and the other side is 1 cm less than twice the shortest side. What is the area of the triangle?

3

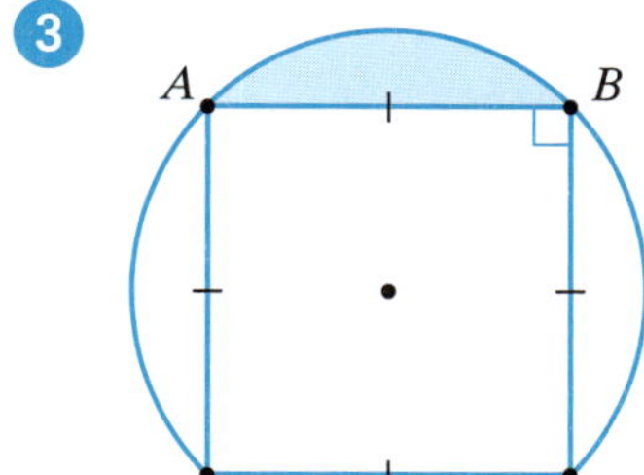

A circle is drawn through the vertices of a square *ABCD*. If $AB = 4$ cm, find the area of the shaded region, in terms of π.

4 A ladder is 2.5 metres long and leans against a wall. The end of the ladder is 70 cm from the base of the wall. If the top of the ladder slips 40 cm down the wall, how far does the end of the ladder move out from the wall?

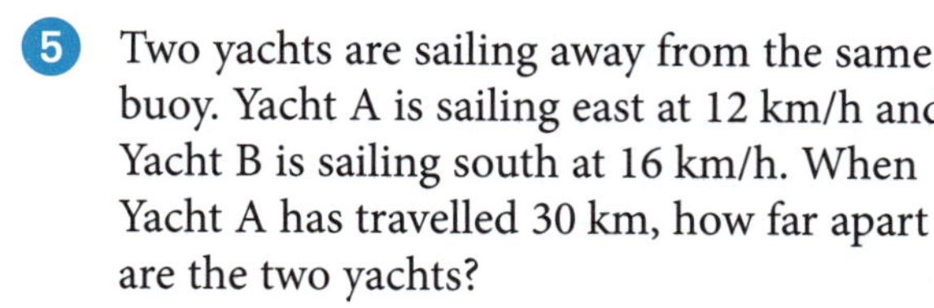

5 Two yachts are sailing away from the same buoy. Yacht A is sailing east at 12 km/h and Yacht B is sailing south at 16 km/h. When Yacht A has travelled 30 km, how far apart are the two yachts?

6 CHALLENGE

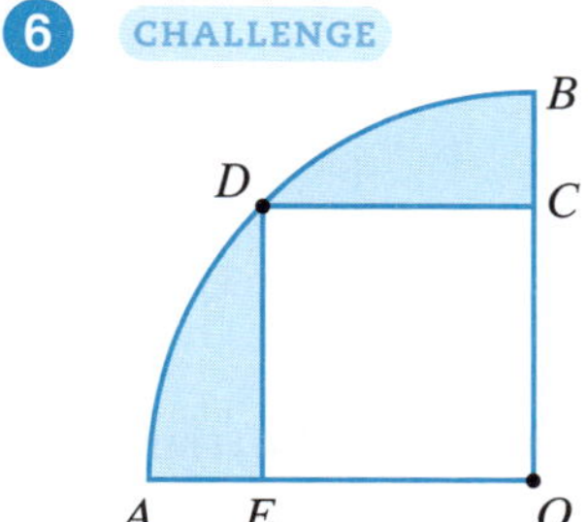

A square *OEDC* sits inside a quadrant *OADB*. If $DC = 5$ cm, find the area of the shaded region, leaving your answer in terms of π.

Answers pages 144–145

KEY SKILL

30 Perimeter of composite shapes

HINTS

- The perimeter is the distance around the outside of a plane shape.
- The circumference is the perimeter of a circle:

  $C = \pi d$ or $C = 2\pi r$,

  where C = circumference, d = diameter,

  r = radius,

  e.g. Find the circumference of a circle with radius 9 cm. Leave your answer correct to 3 significant figures.

  $C = 2 \times \pi \times 9 = 56.4$ (3 sig. figs)

  $\therefore$ the circumference is 56.4 cm.

Examples

A garden is in the shape of a square of side 6 m with one side joining a semicircle. A border is built around the garden at a cost of $9.80/m. Find the total cost of the border to the nearest dollar.

Solution

$$\text{Perimeter} = 6 \times 3 + \frac{1}{2} \times 2 \times \pi \times 3$$
$$= 27.425 \text{ (3 dec. pl.)}$$
$$\text{Cost} = 9.8 \times 27.425$$
$$= 269 \text{ (nearest whole)}$$

$\therefore$ cost will be $269

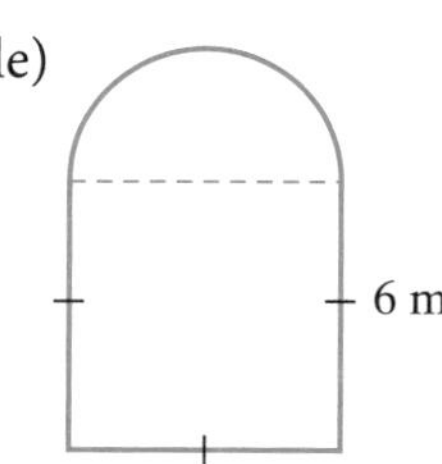

FOCUS on...

1. The question: Asks you to find the cost of the border of the garden.
2. The information: Gives you the dimensions of the square.
3. Your working: Draw a diagram of a square and a semicircle. The diameter of the semicircle is the same as the side of the square.
   The formula for the circumference is $C = 2\pi r$.
   You add the perimeter of 3 sides of the square and half the circumference of a circle.
   $$\text{Perimeter} = 6 \times 3 + \frac{1}{2} \times 2 \times \pi \times 3$$
   $$= 27.424\,777\,96...$$
   $$= 27.425 \text{ (3 dec. pl.)}$$
   You multiply the perimeter by the cost/m.
   $$\text{Cost} = 27.425 \times 9.8$$
   $$= 269 \text{ (nearest whole)}$$
4. Your answer: Make sure you write the correct units:
   The cost will be $269.

Another garden is made from a right triangle with a semicircle attached to the triangle along its hypotenuse. One side of the triangle is 8 m long and the diameter of the semicircle is 10 m. Kendall walks around the outside of the garden at an average speed of 1.4 m/s. How long does it take for Kendall to walk around the garden?

Solution

$$x^2 = 10^2 - 8^2$$
$$= 100 - 64 = 36$$
$$x = 6$$
$$\text{Perimeter} = 6 + 8 + \frac{1}{2} \times 2 \times \pi \times 5$$
$$= 29.71 \text{ (2 dec. pl.)}$$
$$\text{Time} = \text{Distance} \div \text{Speed}$$
$$= 29.71 \div 1.4$$
$$= 21.22 \text{ (2 dec. pl.)}$$

$\therefore$ it takes Kendall 21.22 seconds.

FOCUS on...

1. The question: Asks you to find the time it takes to walk around the outside of the garden.
2. The information: Gives you two sides of the triangle and the shared side with the semicircle.
3. Your working: Draw a diagram of a triangle and a semicircle, where the hypotenuse is the diameter.
   Apply Pythagoras' theorem.
   $$x^2 = 10^2 - 8^2$$
   $$= 100 - 64$$
   $$= 36$$
   $$x = 6 \quad \text{The base is 6 m.}$$
   The formula for the circumference is $C = 2\pi r$.
   You add the two sides of the triangle and half the circumference of a circle.
   $$\text{Perimeter} = 6 + 8 + \frac{1}{2} \times 2 \times \pi \times 5 = 29.71 \text{ (2 dec. pl.)}$$
   To find time, you use $T = \frac{D}{S}$.
   $$T = 29.71 \div 1.4 = 21.22 \text{ (2 dec. pl.)}$$
4. Your answer: Make sure you write the correct units:
   It takes Kendall 21.22 seconds to walk around.

Now try these!

For questions 1 to 5, find the exact perimeters of the following shapes.

1

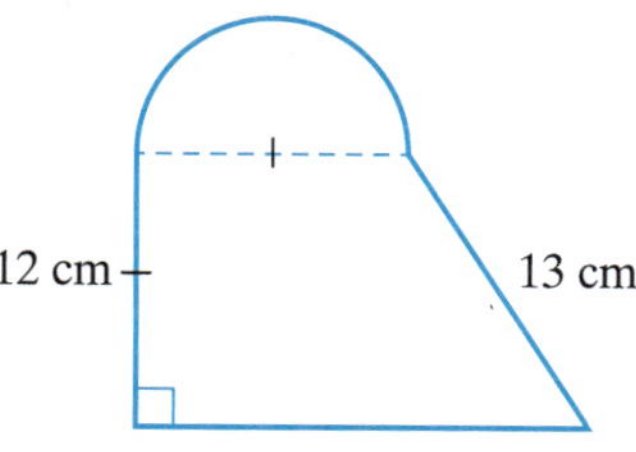

2

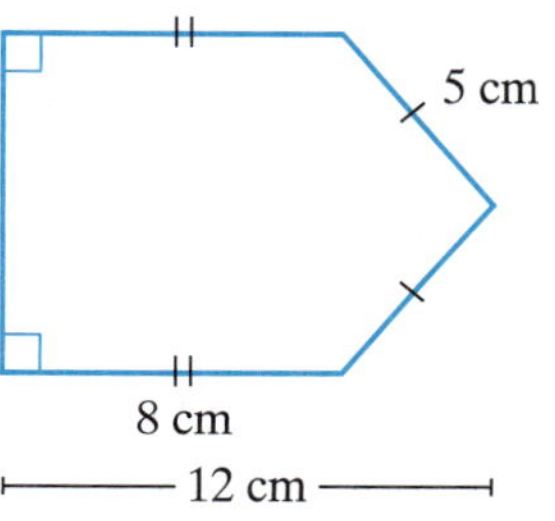

3

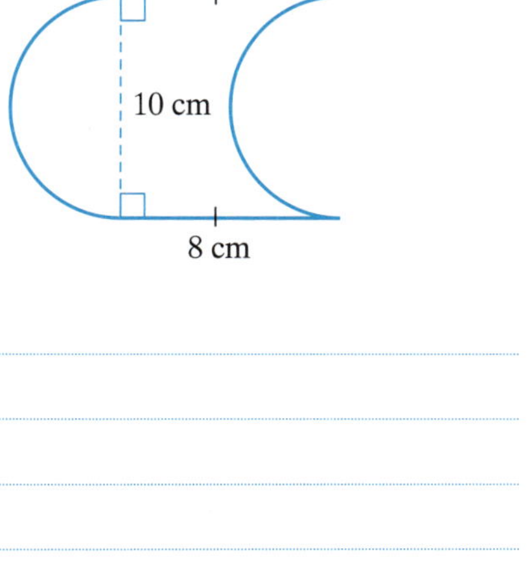

4

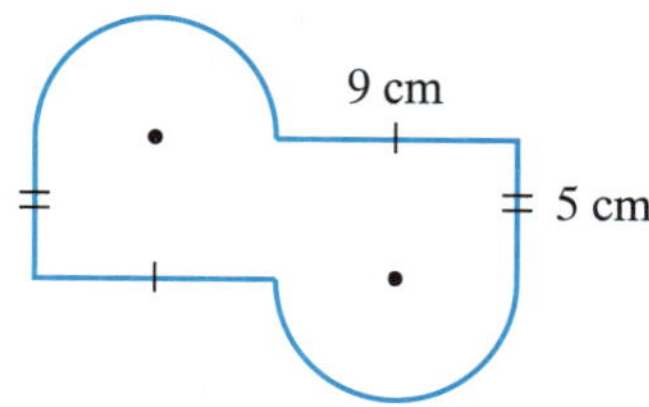

5

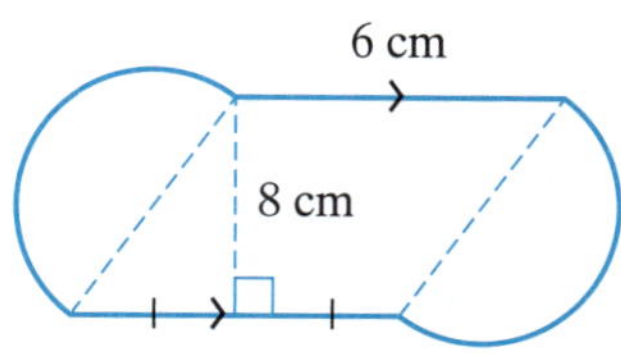

6 CHALLENGE

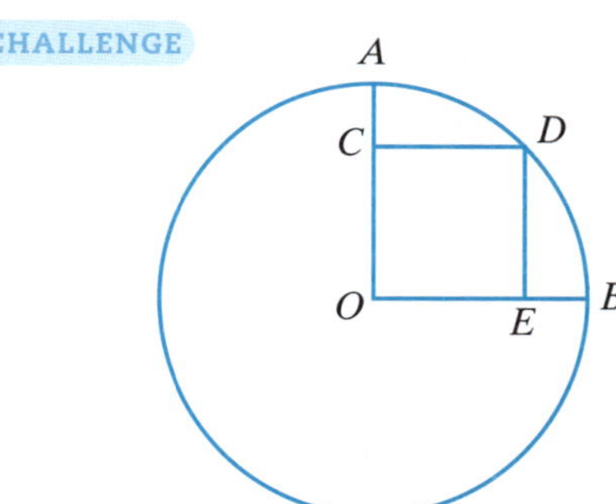

O is the centre of the circle. If the area of square *OCDE* is 2 cm^2, find the exact length of the arc *ADB*?

Answers page 145

KEY SKILL

31 Area of composite shapes

HINTS

- The area is a measure of the space inside a plane shape.
- The formulae are:

Square: $A = s^2$

Rectangle: $A = lb$

Triangle: $A = \frac{1}{2}bh$

Parallelogram: $A = bh$

Trapezium: $A = \frac{1}{2}h(a + b)$

Circle: $A = \pi r^2$

Rhombus/Kite: $A = \frac{1}{2}$product of diagonals

Reminder!

- Read the question carefully to identify what needs to be found.
- Re-read the question when you've finished your answer to make sure that the question has been answered and your solution makes sense.

Examples

Jack has a patio in the shape of a square of side 8 m. He wants to pave an area in the shape of a quadrant and the remaining area is to be a garden. What is the area of the garden, correct to 2 decimal places?

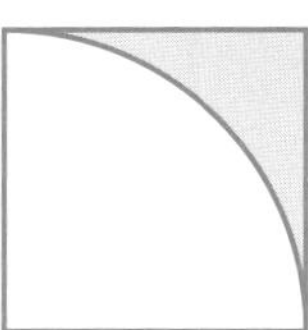

Solution

$$\text{Shaded area} = 8^2 - \frac{1}{4} \times \pi \times 8^2$$
$$= 13.734\,517\,54 \ldots$$
$$= 13.73 \text{ (2 dec. pl.)}$$

$\therefore$ the area is 13.73 m^2.

FOCUS on ...

1. The question: Asks you to find the area of a garden.
2. The information: Gives you side length of a square.
3. Your working: Draw a diagram of a square and a quadrant, and shade in the garden.
   The formula for the area of a circle is $A = \pi r^2$.
   You subtract the area of quadrant (quarter of a circle) from the area of the square.
   $$\text{Area} = 8^2 - \frac{1}{4} \times \pi \times 8^2$$
   $$= 13.734\,517\,54 \ldots$$
   $$= 13.73 \text{ (2 dec. pl.)}$$
4. Your answer: Make sure you write the correct units:
   The area is 13.73 m^2.

A logo is made from a rectangle and three semicircles. The length of the rectangle is 8 cm and the area of the rectangle is 32 cm^2. Find the area of the logo, correct to 3 significant figures.

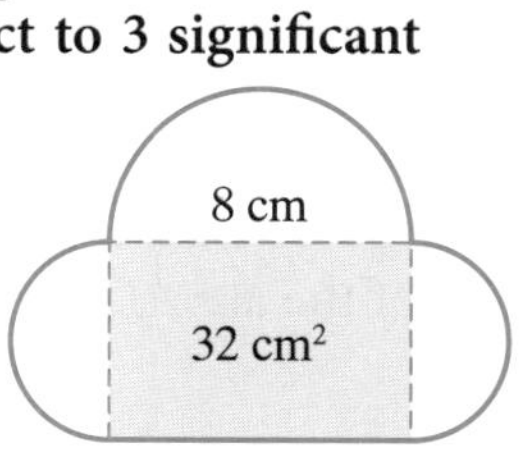

Solution

$$\text{Width of rectangle} = 32 \div 8$$
$$= 4$$
$$\text{Area} = 32 + \frac{1}{2} \times \pi \times 4^2 + \pi \times 2^2$$
$$= 69.699\,111\,84\ldots$$
$$= 69.7 \text{ (3 sig. figs.)}$$

$\therefore$ area is 69.7 cm^2

FOCUS on ...

1. The question: Asks you to find the area of a logo made from three semicircles and a rectangle.
2. The information: Gives you the length and the area of the rectangle.
3. Your working: You divide to find the width of the rectangle: $32 \div 8 = 4$.
   Dimensions of the rectangle are 8 cm and 4 cm.
   The diameter of each small semicircle is 4 cm, so that the radii are 2 cm.
   The diameter of the large semicircle is 8 cm, so that the radius is 4 cm.
   You add the areas of the rectangle, the larger semicircle and the two smaller semicircles.
   $$\text{Area} = 32 + \frac{1}{2} \times \pi \times 4^2 + 2 \times \frac{1}{2} \times \pi \times 2^2$$
   $$= 69.699\,111\,84\ldots$$
   $$= 69.7 \text{ (3 sig. figs.)}$$
4. Your answer: Make sure you write the correct units:
   The area is 69.7 cm^2.

Now try these!

1

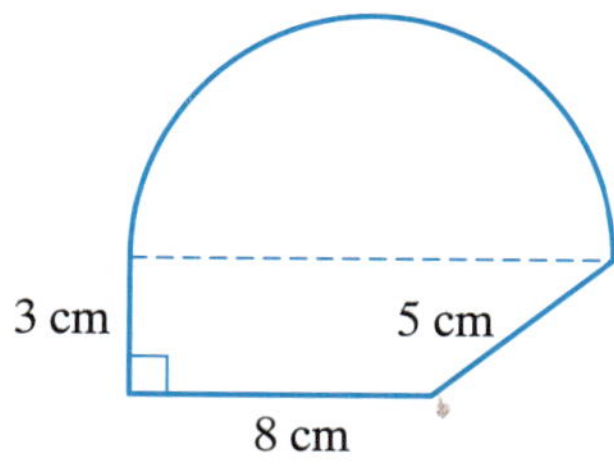

Find the area, in terms of π.

2

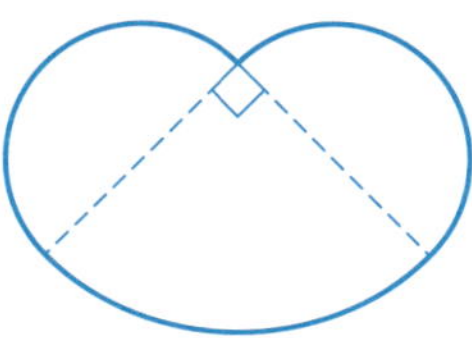

The perimeter of the shape is 3π cm. Express the area in terms of π.

3

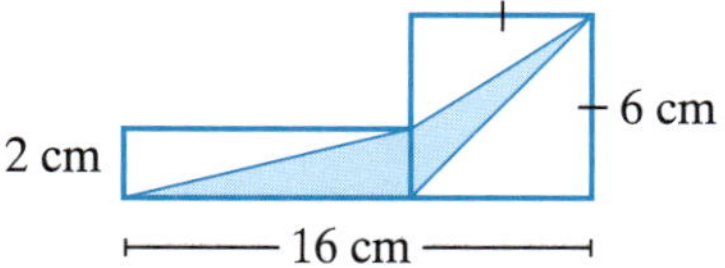

Find the area of the shaded region.

4

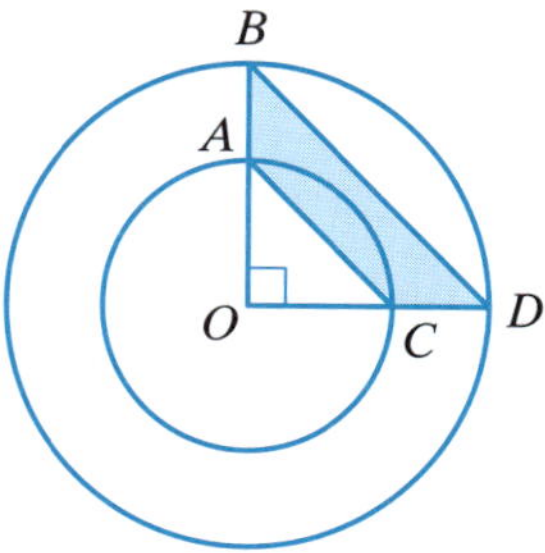

Two concentric circles with centre O where $OC = DC$. If the shaded area is 12 cm^2, find the area of the small circle.

5

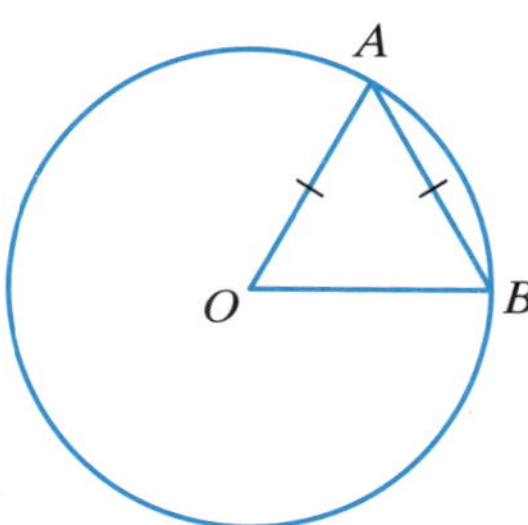

$AO = AB$. The length of arc AB is 2π cm. What is the exact area of the circle?

6 CHALLENGE

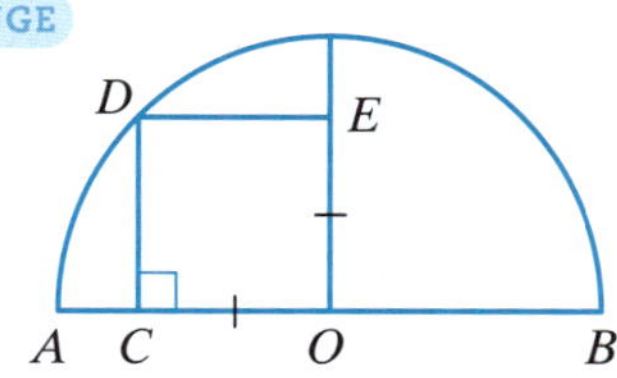

What fraction of the semicircle is $OCDE$?

Answers pages 145–146

KEY SKILL

32 Surface area of prisms

HINTS

- The surface area of a prism is the sum of the areas of each of its sides:
  - cube—6 identical square faces
  - rectangular prism—three pairs of rectangular faces
  - triangular prism—a pair of triangular faces plus three rectangular faces.
- The net of a prism is the pattern used to make the solid.

Examples

Miriam has a square cake pan measuring 15 cm on each side and 8 cm deep. She needs to coat the inside of the pan with baking paper. What area of paper will she need?

Solution

$$SA = 15^2 + 4 \times 15 \times 8$$
$$= 705$$

$\therefore$ 705 cm^2 of paper is needed.

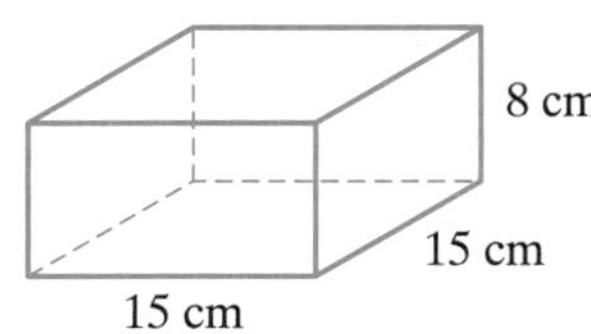

FOCUS on ...

1. The question: Asks you to find the surface area of an open square-based prism.
2. The information: Gives you the dimensions of the prism.
3. Your working: You need to total the areas of five faces.
   Surface area = $15 \times 15 + 4 \times 15 \times 8$
   $= 705$
4. Your answer: Make sure you write the correct units:
   Miriam needs 705 cm^2 of paper.

A block of cheese in the shape of a triangular prism is wrapped. What is the minimum amount of packaging required?

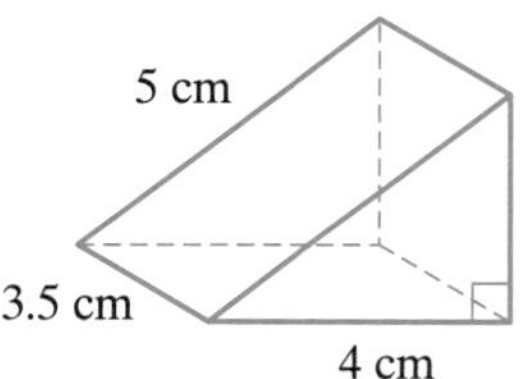

Solution

Let the height of the triangular face be x.
Using Pythagoras:

$$x^2 = 5^2 - 4^2$$
$$= 25 - 16$$
$$= 9$$
$$x = 3$$

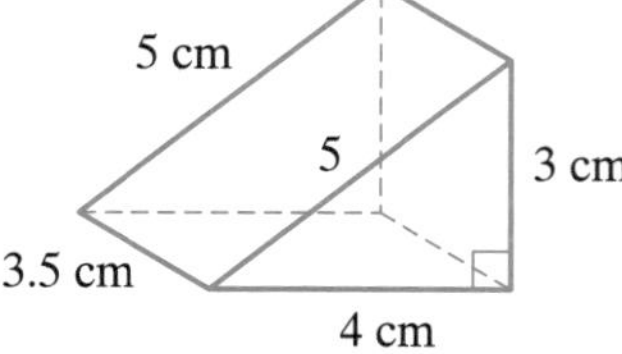

$$SA = 2 \times \frac{1}{2} \times 4 \times 3 + 4 \times 3.5 + 3.5 \times 3 + 5 \times 3.5$$
$$= 54$$

$\therefore$ the minimum amount of packaging needed is 54 cm^2.

FOCUS on ...

1. The question: Asks you to find the surface area of a triangular prism.
2. The information: Gives you a diagram with the dimensions of the prism.
3. Your working: Apply Pythagoras' theorem.
   $x^2 = 5^2 - 4^2$
   $= 25 - 16$
   $= 9$
   $x = 3$
   You need to total the areas of five faces.
   Surface area = $2 \times \frac{1}{2} \times 4 \times 3 + 4 \times 3.5 + 3.5 \times 3 + 5 \times 3.5$
   $= 54$
4. Your answer: Make sure you write the correct units:
   The minimum amount of packaging needed is 54 cm^2.

Now try these!

1. An open box has a square base with sides 16 cm and height of 10 cm. Find the surface area of the box.

2. A prism is 20 cm high and its base is a triangle with sides 5 cm, 12 cm and 13 cm. What is the surface area of the prism?

3. Fiona made a rectangular fence around her roses to protect them from her dog Otis. The fence was 4 metres long, 3 metres wide and 1.6 metres high. How much material was in the fence?

4. A rectangular prism has a surface area of 472 cm^2. The prism has a length of 12 cm and a width of 7 cm. What is the height of the prism?

5. 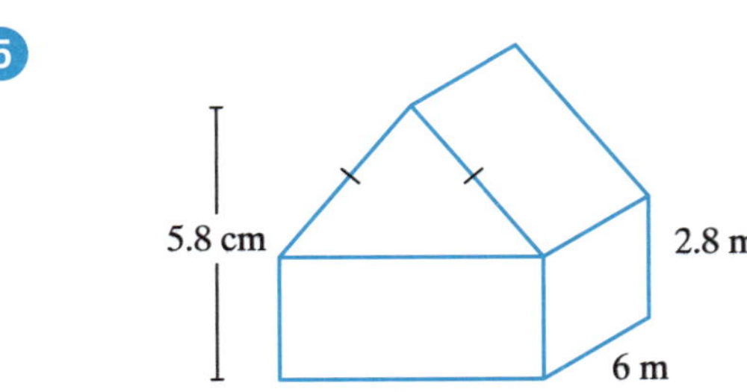

The four walls and the roof of a barn are to be painted with two coats using paint that covers at the rate of 8 m^2 per litre. How much paint will be required?

6. CHALLENGE The total surface area of a closed box is 288 cm^2. The box is 9 cm high and has a square base. Find the length of a side of its base.

Answers pages 146–147

KEY SKILL

33 Surface area of cylinders A

HINTS

- The net of a cylinder is comprised of two circles and a rectangle.
- There are two formulae to remember:
  Closed cylinder: SA $= 2\pi r^2 + 2\pi rh$
  Open cylinder: CSA $= 2\pi rh$
  (CSA: Curved surface area)

e.g. Find the surface area of a closed cylinder of radius 8 cm and height 10 cm, correct to 2 decimal places.

SA $= 2\pi r^2 + 2\pi rh$
$= 2 \times \pi \times 8^2 + 2 \times \pi \times 8 \times 10$
$= 904.78$ (2 dec. pl.)
$\therefore$ 904.78 cm^2.

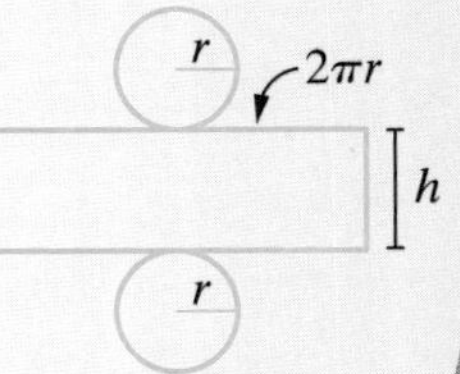

Examples

A cylindrical storage tank is 12 m high and has a diameter of 3.5 m. The tank has a base and a top and is made from stainless steel sheeting which costs $54/m^2. What was the total cost of sheeting, to the nearest dollar?

Solution

radius $= 1.75$

SA $= 2\pi r^2 + 2\pi rh$
$= 2 \times \pi \times 1.75^2 + 2 \times \pi \times 1.75 \times 12$
$= 151.189\ldots$
$= 151.19$ (2 dec. pl.)

Cost $= 151.19 \times 54$
$= 8164$ (nearest whole)

$\therefore$ the cost was $8164.

12 m
3.5 m

FOCUS on …

1. The question: Asks you to find the total cost of the sheeting.
2. The information: Gives you the dimensions of the cylinder and the cost per square metre.
3. Your working: You divide the diameter by 2 to find the radius: $3.5 \div 2 = 1.75$.
   You need the surface area of a cylinder formula:
   SA $= 2\pi r^2 + 2\pi rh$
   $= 2 \times \pi \times 1.75^2 + 2 \times \pi \times 1.75 \times 12$
   $= 151.189\,1465\ldots$
   $= 151.19$ (2 dec. pl.)
   Cost $= 151.19 \times 54$
   $= 8164$ (nearest whole)
4. Your answer: Make sure you write the correct units: The total cost is $8164.

An open cylindrical pipe has an outer diameter of 6.8 cm and is 0.3 cm thick. If the pipe is 2 metres in length, find the total surface area of the pipe in square centimetres, correct to 3 significant figures.

Solution

3.4 cm
0.3 cm
O
3.1 cm

2 m = 200 cm
Outer radius = 3.4 cm
Inner radius = 3.1 cm

Outer SA $= 2\pi rh$
$= 2 \times \pi \times 3.4 \times 200$
$= 4272.566\ldots$

Inner SA $= 2 \times \pi \times 3.1 \times 200$
$= 3895.574\ldots$

SA of 2 ends $= 2 \times (\pi \times 3.4^2 - \pi \times 3.1^2)$
$= 12.252\ldots$

Total SA $= 8180$ (3 sig. figs)

$\therefore$ the total surface area is 8180 cm^2.

FOCUS on …

1. The question: Asks you to find the total surface area.
2. The information: Gives you the outer diameter, the thickness and the length of the pipe.
3. Your working: Draw a diagram to show the area of the cross-section.
   You divide the outer diameter by 2 to find the outer radius: $6.8 \div 2 = 3.4$. As the pipe is 0.3 thick, to find the inner radius you subtract: $3.4 - 0.3 = 3.1$.
   You use formula for curved surface area: $2\pi rh$
   Outer Surface A $= 2 \times \pi \times 3.4 \times 200$
   $= 4272.566\ldots$
   $= 4272.57$ (2 dec. pl.)
   Inner Surface A $= 2 \times \pi \times 3.1 \times 200$
   $= 3895.574\ldots$
   $= 3895.57$ (2 dec. pl.)
   SA of 2 ends $= 2 \times (\pi \times 3.4^2 - \pi \times 3.1^2)$
   $= 12.252\ldots$
   $= 12.25$ (2 dec. pl.)
   Total $= 4272.57 + 3895.57 + 12.25 = 8180$ (3 sig. fig.)
4. Your answer: Make sure you write the correct units: The total surface area is 8180 cm^2.

Now try these!

1. Find the cost of covering a closed cylinder with material at $18.40/m^2 if the cylinder has a radius of 3.2 metres and a height of 9.5 metres. Give your answer to the nearest dollar.

2. An open cylindrical pipe has an outer radius of 4 cm and an inner radius of 3 cm. If the pipe is 3 metres long, what is the total surface area in square centimetres, correct to the nearest square centimetre?

3. A lawn roller is 0.9 metres wide and 0.8 metres high.

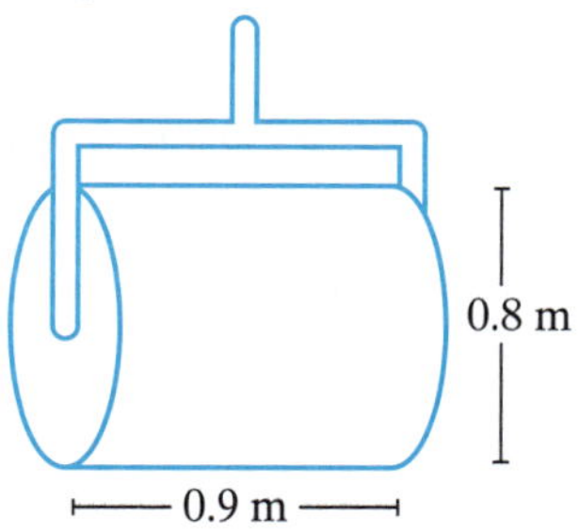

What area will be rolled in 100 revolutions of the roller, to the nearest square metre?

4. Find the area of a label on a can of beans with diameter 7.5 cm and height 10.5 cm, to 3 significant figures.

5. 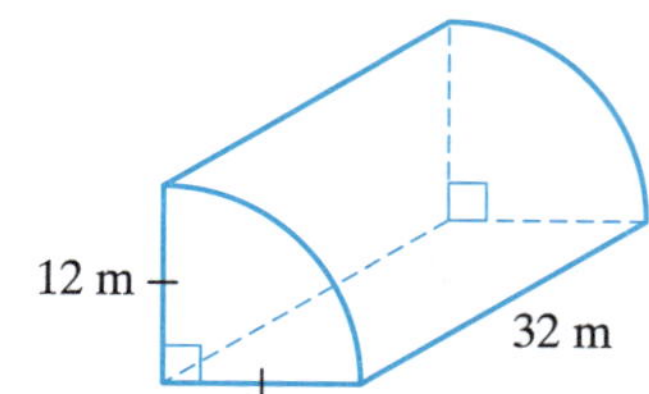

The shape is to be covered with sheet metal at a cost of $15/m^2. What is the total cost, to the nearest 1000 dollars?

6. CHALLENGE

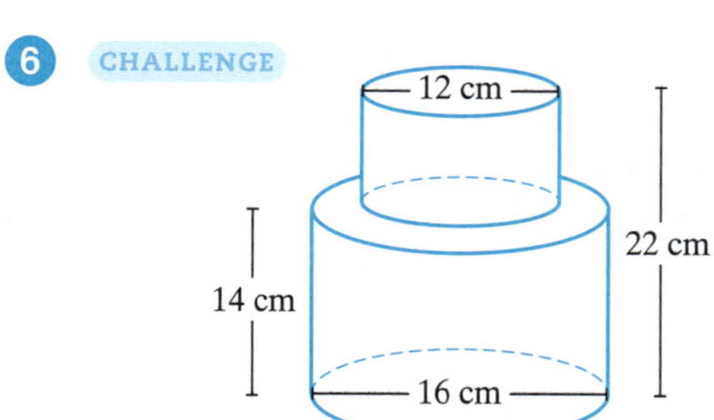

A closed cylinder is glued to the top of another. What is the total surface area to be sprayed, in terms of π?

Answers page 147

KEY SKILL

34 Surface area of cylinders B

HINTS

- Make sure you know the **Surface area of cylinders Hints** from Key Skill 33 on page 82.
- The subject of a formula is the single variable (usually on the left-hand side) that everything else is equal to. 'To change the subject' means to rearrange the formula so that another variable is the subject of the formula.

Examples

A can with a height of 8 cm has a curved surface area of 48π cm^2. What is the total surface area of the can, in terms of π?

Solution

$$\begin{aligned} CSA &= 2\pi rh \\ 48\pi &= 2\pi r \times 8 \\ 16r &= 48 \\ r &= 3 \\ SA &= 2\pi r^2 + 2\pi rh \\ &= 2\pi \times 3^2 + 48\pi \\ &= 66\pi \end{aligned}$$

$\therefore$ the area is 66π cm^2.

FOCUS on...

1. The question: Asks you to find total surface area.
2. The information: Gives you the height of the cylinder and the curved surface area.
3. Your working: You need to first find the radius using CSA = $2\pi rh$ and change the subject to r.
   $$\begin{aligned} 48\pi &= 2\pi r \times 8 \\ 16r &= 48 \\ \frac{16r}{16} &= \frac{48}{16} \\ r &= 3 \end{aligned}$$
   Subs. $r = 3$ and $h = 8$ in SA = $2\pi r^2 + 2\pi rh$
   $$SA = 2\pi \times 3^2 + 48\pi = 66\pi$$
4. Your answer: Make sure you write the correct units: The total surface area is 66π cm^2.

A cylindrical tank has a height of 4 metres and a total surface area of 10π m^2. If the curved surface is to be painted at the rate of 2.3 m^2/L, how much paint will be required, to the nearest litre?

Solution

$$\begin{aligned} SA &= 2\pi r^2 + 2\pi rh \\ 10\pi &= 2\pi r^2 + 2\pi r \times 4 \\ 2\pi r^2 + 8\pi r &= 10\pi \\ r^2 + 4r - 5 &= 0 \\ (r + 5)(r - 1) &= 0 \\ r &= -5, 1 \\ r &= 1 \text{ (as } r > 0) \\ CSA &= 2\pi rh \\ &= 2 \times \pi \times 1 \times 4 = 8\pi \\ \text{Paint} = 8\pi \div 2.3 &= 10.927\ldots \\ &= 11 \text{ (nearest whole)} \end{aligned}$$

$\therefore$ needs 11 L of paint.

FOCUS on...

1. The question: Asks you to find the amount of paint required.
2. The information: Gives you the height of the cylinder, the total surface area and the rate of coverage of the paint.
3. Your working: You need to first find the radius using SA = $2\pi r^2 + 2\pi rh$ and change the subject to r.
   $$\begin{aligned} 10\pi &= 2\pi r^2 + 2\pi r \times 4 \\ 2\pi r^2 + 8\pi r &= 10\pi \\ r^2 + 4r - 5 &= 0 \\ (r + 5)(r - 1) &= 0 \\ r &= -5, 1 \\ r &= 1 \quad (\text{as } r > 0) \end{aligned}$$
   Subs. $r = 1$ and $h = 4$ in CSA = $2\pi rh$.
   $$\begin{aligned} CSA &= 2 \times \pi \times 1 \times 4 \\ &= 8\pi \end{aligned}$$
   You now divide the area by 2.3.
   $$\begin{aligned} \text{Paint} &= 8\pi \div 2.3 = 10.927\ldots \\ &= 11 \text{ (nearest whole)} \end{aligned}$$
4. Your answer: Make sure you write the correct units: 11 litres of paint is required.

Now try these!

1. The curved surface area of a cylinder with height 12 cm is 144π cm^2. What is the total surface area of the cylinder, in terms of π?

2. The total surface area of a cylinder with a height of 10 m is 288π m^2. Find the cost of covering the curved surface area of the cylinder at \$9.40/m^2. Leave your answer to the nearest dollar.

3. A cylinder with a radius of 9 cm has a surface area of 450π. What is the height of the cylinder?

4. The total surface area of a cylinder is 128π cm^2. If the height of the cylinder is 12 cm, what is the curved surface area of the cylinder? Give your answer correct to 2 significant figures.

5. 

The curved surface area is 160π cm^2. What is the area of the base of the cylinder, correct to one decimal place?

6. CHALLENGE

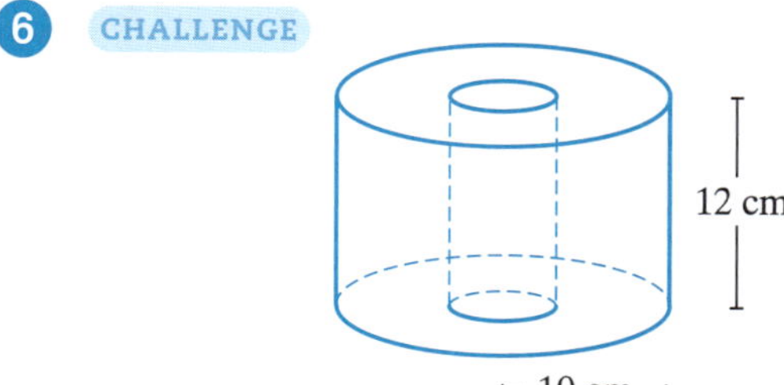

The hollow pipe has a height of 12 cm and the radius of the outer cylinder is 10 cm. The total surface area of the figure is 504π cm^2. What are the possible radii of the inside cylinder?

Answers pages 147–148

REVISION TEST 9 Level of difficulty—Average

1

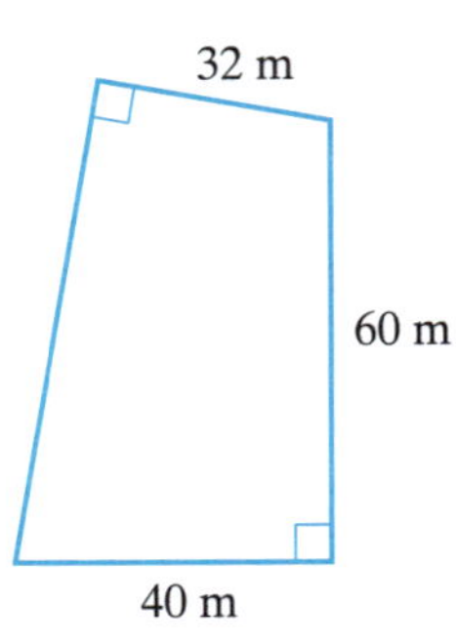

The diagram shows a paddock. If fencing costs \$8.25/metre, find the cost of fencing the paddock, to the nearest dollar.

2 Two jets leave from the same airport. One jet leaves at 3 pm and travels north at an average speed of 1000 km/h. The other jet leaves the airport at 4:15 pm and travels west at an average speed of 1200 km/h. How far apart are the jets at 5:00 pm, to the nearest 100 kilometres?

3 A rectangular prism has a surface area of 416 cm^2. The prism has a length of 10 cm and a height of 6 cm. What is the width of the prism?

4

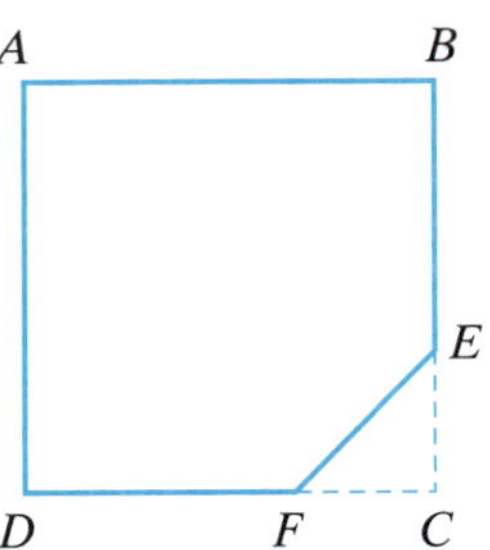

The perimeter of the square $ABCD$ is 80 cm. If $EC = FC$ and $EF = 4$ cm, what is the area of pentagon $ABEFD$?

5

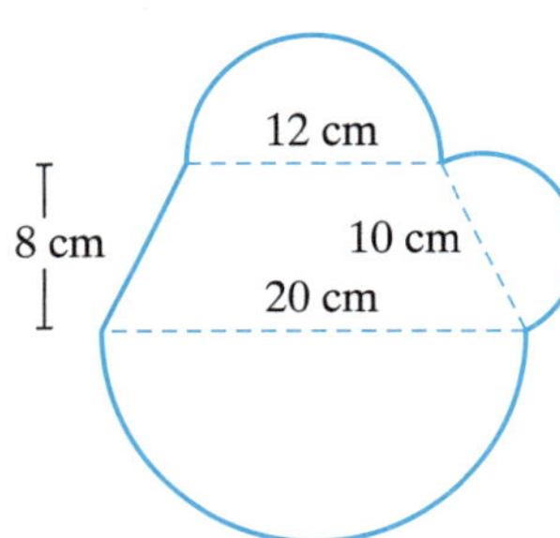

A shape is comprised of a trapezium and three semicircles. Find the exact area of the shape.

6 The curved surface area of a cylinder with height 18 cm is 540π cm^2. What is the total surface area of the cylinder, in terms of π?

Answers page 148

REVISION TEST  Level of difficulty—Challenging

1. Meredith uses two pieces of string of the same length to make a square and a circle. What is the ratio of the area of the square to the area of the circle?

2. 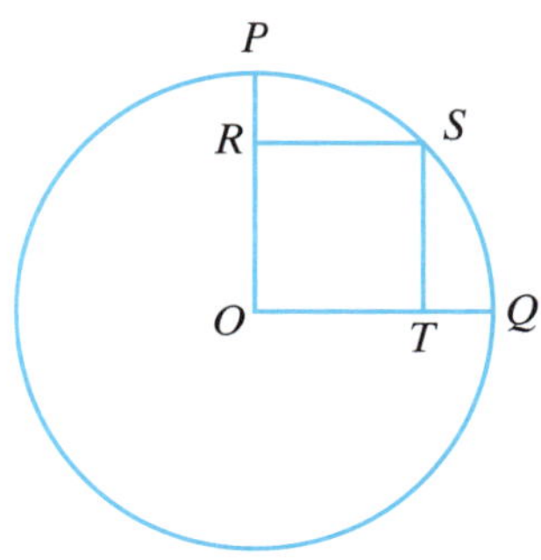

O is the centre of the circle. If the area of square $ORST$ is 4 cm^2, find the exact length of the arc SQ?

3. Jack melted three metal cubes to form a larger cube. The three smaller cubes had sides of 3 cm, 4 cm and 5 cm each. What is the ratio of the surface area of the large cube to the total surface area of the three smaller cubes?

4. A cabling company places four round cables in a round duct. If the diameter of each cable is 4 mm, what would be the exact diameter of the duct?

5. The total surface area of a closed box is 800 cm^2. The box is 15 cm high and has a square base. Find the volume of the box.

6. 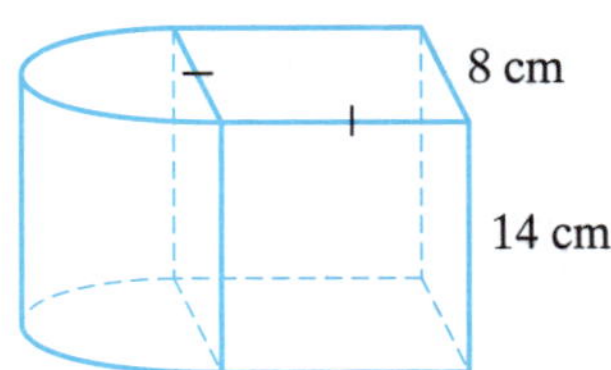

Find the surface area of the solid, in terms of π.

Answers pages 148–149

KEY SKILL

35 Volume of prisms

HINTS

- The volume of a prism is found by using $V = Ah$
  where V = volume
  A = area of the base
  h = perpendicular height

e.g. A hexagonal prism of height 5 cm has a base area of 12 cm^2. What is its volume?

$V = 12 \times 5$
$= 60$

$\therefore$ volume is 60 cm^3.

Examples

A block of metal in the shape of a trapezoidal prism is made from an alloy of mass 11g/cm^3. What is the mass of the block, in kilograms?

Solution

$A = \frac{1}{2}h(a + b)$
$= \frac{1}{2} \times 7(4 + 12)$
$= 56$
$V = Ah$
$= 56 \times 6$
$= 336$
Mass $= 336 \times 11$
$= 3696$
$\therefore$ the mass is 3.696 kg.

4 cm
7 cm
6 cm
12 cm

FOCUS on...

1. The question: Asks you to find the mass of the prism.
2. The information: Gives you a diagram of a trapezoidal prism with dimensions, and the mass/cm^3 of the alloy.
3. Your working: You need to first find the area of the trapezium using $A = \frac{1}{2}h(a + b)$.
   $A = \frac{1}{2} \times 7(4 + 12) = 56$
   Now subs. $A = 56$ and $h = 6$ in the formula $V = Ah$.
   $V = 56 \times 6 = 336$
   You now multiply the volume by 11 to find the mass.
   Mass $= 336 \times 11 = 3696$
   Change to kilograms by dividing by 1000.
   $3696 \div 1000 = 3.696$
4. Your answer: Make sure you write the correct units: The mass is 3.696 kg.

A wedge has a volume of 960 cm^3. What is the surface area of the wedge?

Solution

Let base be x:
$x^2 = 20^2 - 12^2$
$= 400 - 144$
$= 256$
$x = 16$
$V = Ah$
$960 = \frac{1}{2} \times 16 \times 12 \times h$
$96h = 960$
$h = 10$
$SA = 2 \times \frac{1}{2} \times 16 \times 12 + 16 \times 10 + 20 \times 10 + 12 \times 10$
$= 672$
$\therefore$ surface area is 672 cm^2.

20 cm
12 cm
x

FOCUS on...

1. The question: Asks you to find the surface area.
2. The information: Gives you a diagram of a triangular prism with height and hypotenuse, and the volume.
3. Your working: Apply Pythagoras' theorem:
   $x^2 = 20^2 - 12^2$
   $= 400 - 144$
   $= 256$
   $x = 16$
   Area of triangle $= \frac{1}{2} \times 16 \times 12 = 96$
   Now subs. $V = 960$ and $A = 96$ in the formula $V = Ah$.
   $960 = 96 \times h$
   $h = \frac{960}{96}$
   $= 10$
   You now add the areas of 5 faces:
   $SA = 2 \times \frac{1}{2} \times 16 \times 12 + 16 \times 10 + 20 \times 10 + 12 \times 10$
   $= 672$
4. Your answer: Make sure you write the correct units: The surface area is 672 cm^2.

Now try these!

1

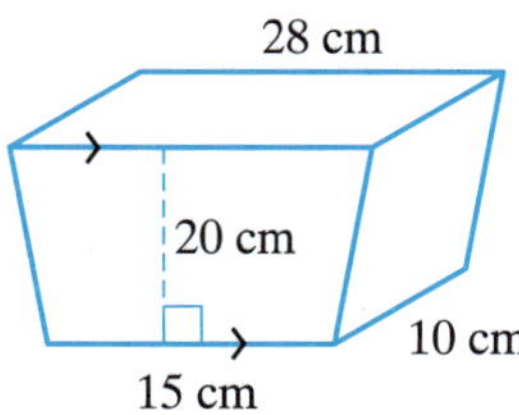

Find the mass, in kilograms, of this prism if it is made from metal with a mass of 16 g/cm^3.

2

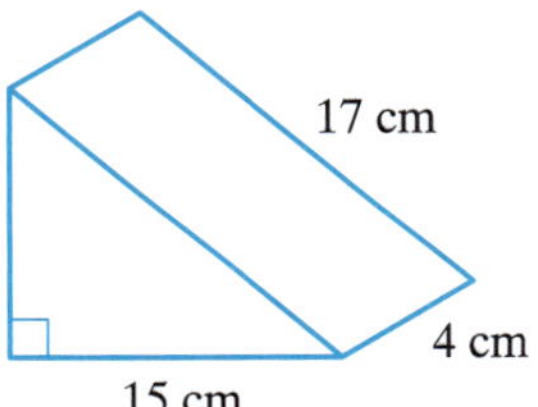

Find the volume.

3 A cube with surface area of 3456 cm^2 is half filled with sand. The sand is then poured into a rectangular prism with length of 18 cm and width 12 cm. How deep is the sand in the prism?

4 The length of a rectangular prism is 3 cm longer than its width and its height is 2 cm. If the volume is 36 cm^3, what is its surface area?

5

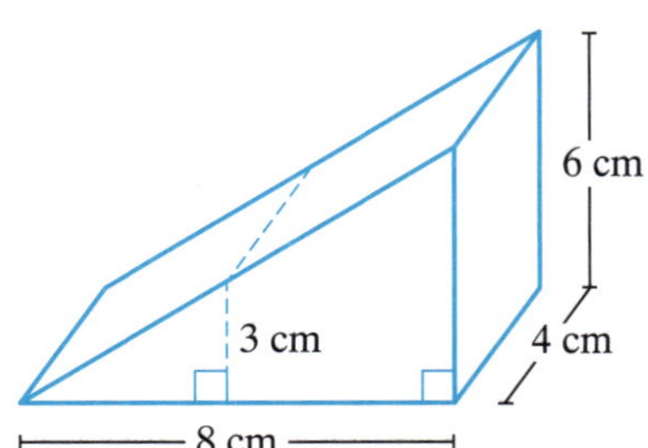

A wooden wedge is jammed under a door with a 3-cm distance between the bottom of the door and the floor. What percentage of the volume of the wedge can be seen?

6 CHALLENGE

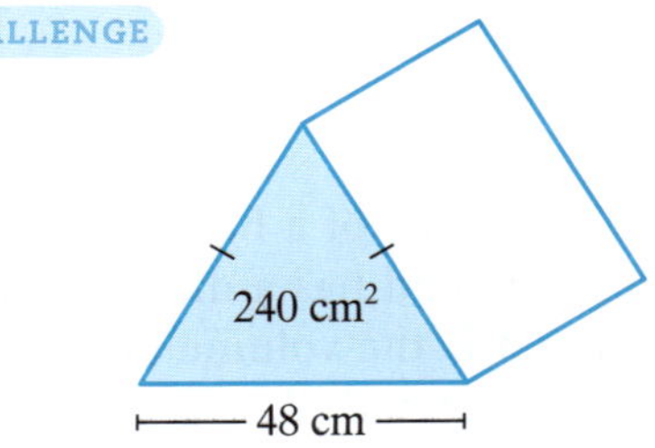

If the volume is 6000 cm^3, find the surface area.

Answers page 149

KEY SKILL

36 Volume of cylinders A

HINTS

- The volume of a cylinder is also $V = Ah$, where $A = \pi r^2$. This is expressed as $V = \pi r^2 h$, where
  V = volume
  r = radius
  h = height

  e.g. Find V, if $V = \pi r^2 h$ and $r = 4$, $h = 5$. Leave answer correct to 2 decimal places.

  $$V = \pi \times 4^2 \times 5$$
  $$= 251.33 \text{ (2 dec. pl.)}$$

Reminder!

- Read the question carefully to identify what needs to be found.
- Re-read the question when you've finished your answer to make sure that the question has been answered and your solution makes sense.

Examples

A drum has a diameter of 90 cm and a height of 1.2 m. Find the volume of the drum in cubic metres, correct to 3 significant figures.

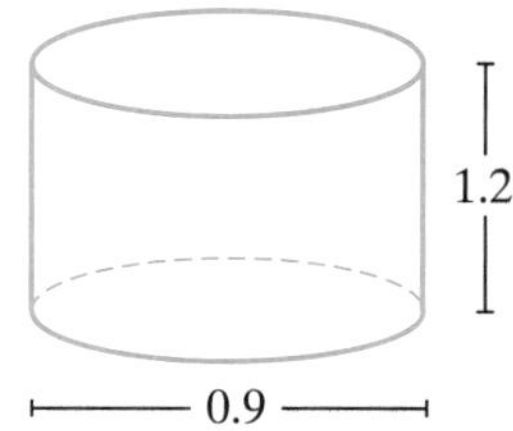

Solution

diameter: 0.9 m,
∴ radius = 0.45 m
height: 1.2 m

$$V = \pi \times 0.45^2 \times 1.2$$
$$= 0.763\,407\,014\ldots$$
$$= 0.763 \text{ (3 significant figures)}$$

∴ the volume is 0.763 m^3.

FOCUS on ...

1. The question: Asks you to find the volume of the cylinder.
2. The information: Gives you the diameter and the height of the cylinder.
3. Your working: Divide the diameter by 2 to find the radius: 90 ÷ 2 = 45.
   Change this to metres by dividing by 100.
   45 ÷ 100 = 0.45.
   Now subs. $r = 0.45$ and $h = 1.2$ in $V = \pi r^2 h$.
   $$V = \pi \times 0.45^2 \times 1.2$$
   $$= 0.763\,407\,014\ldots$$
   $$= 0.763 \text{ (3 sig. figs)}$$
4. Your answer: Make sure you write the correct units: The volume is 0.763 m^3.

A trough in the shape of a half-cylinder is 3.4 metres long and has a diameter of 0.8 metres. What is the volume of the trough, correct to three decimal places?

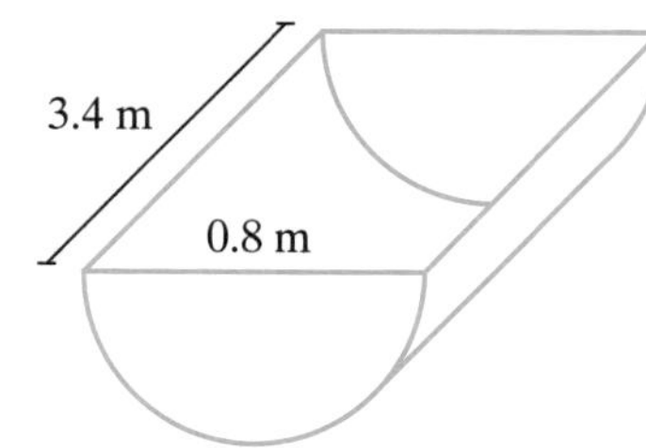

Solution

diameter: 0.8 m
∴ radius = 0.4 m
height: 3.4 m

$$V = \frac{1}{2} \times \pi \times 0.4^2 \times 3.4$$
$$= 0.855 \text{ (3 dec. pl.)}$$

∴ the volume is 0.855 m^3.

FOCUS on ...

1. The question: Asks you to find the volume of half a cylinder.
2. The information: Gives you the diameter and the height of the half-cylinder.
3. Your working: Divide the diameter by 2 to find the radius: 0.8 ÷ 2 = 0.4.
   Now subs. $r = 0.4$ and $h = 3.4$ in $V = \frac{1}{2}\pi r^2 h$.
   $$V = \frac{1}{2} \times \pi \times 0.4^2 \times 3.4$$
   $$= 0.854\,513\,201\ldots$$
   $$= 0.855 \text{ (3 dec. pl.)}$$
4. Your answer: Make sure you write the correct units: The volume is 0.855 m^3.

Now try these!

1. A cylinder has a diameter of 80 cm and a height of 2.3 m. Find the volume of the cylinder, in cubic metres, correct to 2 decimal places.

For questions 2 to 5, find the volume, correct to 2 decimal places.

2.

2.1 m
30 cm

3.

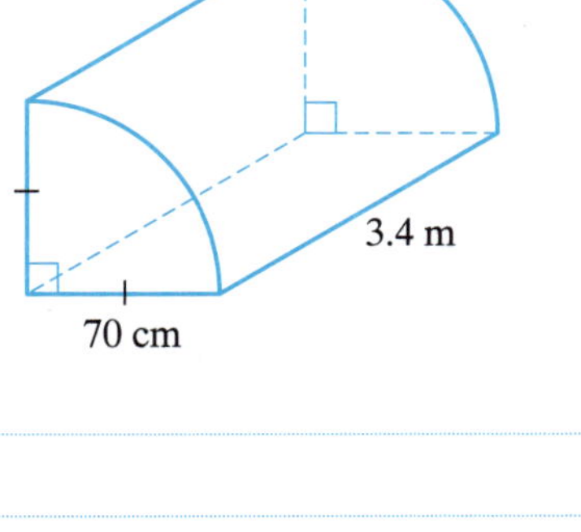

4.

16 cm
10 cm

5.

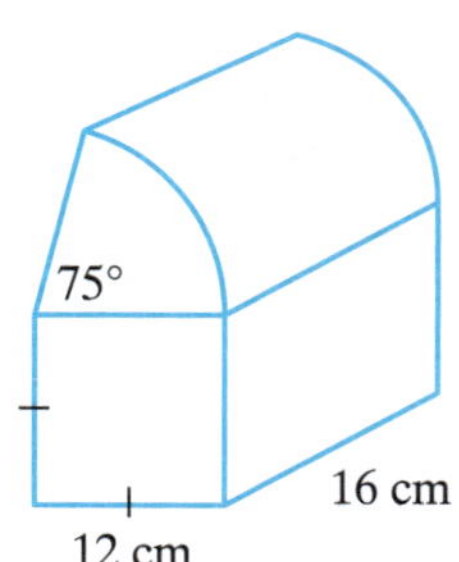

6. CHALLENGE

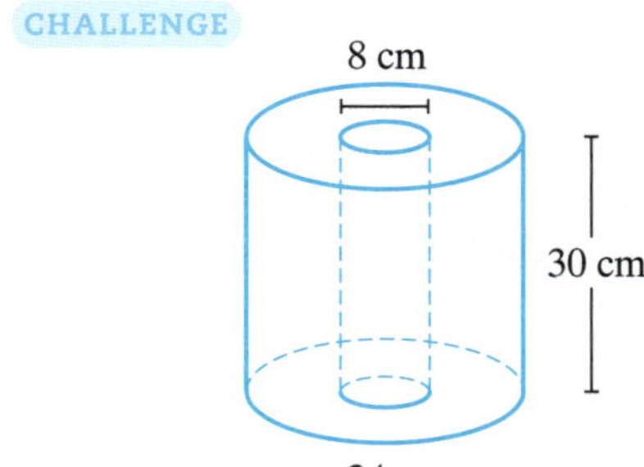

Find the mass, in kilograms, of the shape if it is made of metal with a mass of 3.2 g/cm^3.

Answers page 150

KEY SKILL

37 Volume of cylinders B

HINTS

- Make sure you know the **Volume of cylinders A Hints** from Key Skill 36 on page 90.
- The subject of a formula is the single variable (usually on the left-hand side), that everything else is equal to. 'To change the subject' means to rearrange the formula so that another variable is the subject of the formula.

Examples

The height of a cylinder is the same as the circumference of its base. What is the volume of the cylinder if the height is 26 cm? Leave your answer correct to 3 significant figures.

Solution

$$C = 2\pi r$$
$$26 = 2\pi r$$
$$\therefore \quad r = \frac{13}{\pi}$$
$$V = \pi r^2 h$$
$$= \pi \times \left(\frac{13}{\pi}\right)^2 \times 26$$
$$= 1398.653\,64\ldots$$
$$= 1400 \text{ (3 sig. figs)}$$

$\therefore$ the volume is 1400 cm^3.

FOCUS on...

1. The question: Asks you to find the volume of a cylinder.
2. The information: Gives you the height and the circumference of the circular base.
3. Your working: You need to find the radius by substituting C = 26 in the formula $C = 2\pi r$.
   $$26 = 2\pi r$$
   $$r = \frac{13}{\pi}$$
   To find the volume you substitute $r = \frac{13}{\pi}$ and $h = 26$ in $V = \pi r^2 h$.
   $$V = \pi \times \left(\frac{13}{\pi}\right)^2 \times 26$$
   $$= 1398.653\,64\ldots$$
   $$= 1400 \text{ (3 sig. figs)}$$
4. Your answer: Make sure you write the correct units: The volume is 1400 cm^3.

The total surface area of a cylinder with a radius of 12 cm is 480π cm^2. What is the volume of the cylinder, in terms of π?

Solution

$$SA = 2\pi r^2 + 2\pi rh$$
$$480\pi = 2 \times \pi \times 12^2 + 2 \times \pi \times 12 \times h$$
$$480\pi = 288\pi + 24\pi h$$
$$24\pi h = 480\pi - 288\pi$$
$$24\pi h = 192\pi$$
$$h = \frac{192\pi}{24\pi}$$
$$= 8$$
$$V = \pi r^2 h$$
$$= \pi \times 12^2 \times 8$$
$$= 1152\pi$$

$\therefore$ the volume is 1152π cm^3.

FOCUS on...

1. The question: Asks you to find the volume of a cylinder.
2. The information: Gives you the radius and the total surface area of a cylinder.
3. Your working: You need to find the height by substituting $r = 12$ and $A = 480\pi$ in the formula $A = 2\pi r^2 + 2\pi rh$.
   $$480\pi = 2 \times \pi \times 12^2 + 2 \times \pi \times 12 \times h$$
   $$480\pi = 288\pi + 24\pi h$$
   $$24\pi h = 192\pi$$
   $$\frac{24\pi h}{24\pi} = \frac{192\pi}{24\pi}$$
   $$h = 8$$
   You need to find the volume by substituting $r = 12$ and $h = 8$ in $V = \pi r^2 h$.
   $$V = \pi \times 12^2 \times 8$$
   $$= 1152\pi$$
4. Your answer: Make sure you write the correct units: The volume is 1152π cm^3.

Now try these!

1 The height of a cylinder is the same as the circumference of its base. What is the volume of the cylinder if the height is 18 cm? Leave your answer in terms of π.

2

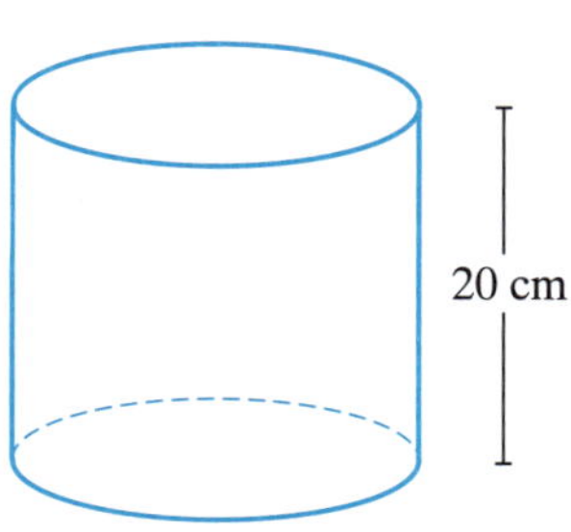

A cylinder has a height of 20 cm and a total surface area of 600π cm^2. Find the volume of the cylinder, in terms of π.

3 Find the exact volume of a cylinder when its curved surface area is 432 cm^2 and the circumference of its base is 48 cm.

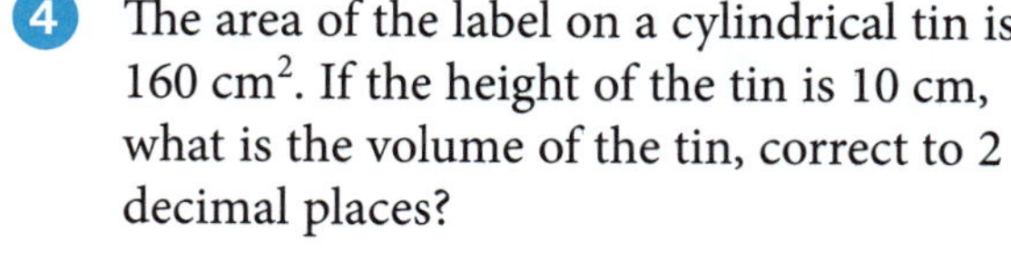

4 The area of the label on a cylindrical tin is 160 cm^2. If the height of the tin is 10 cm, what is the volume of the tin, correct to 2 decimal places?

5

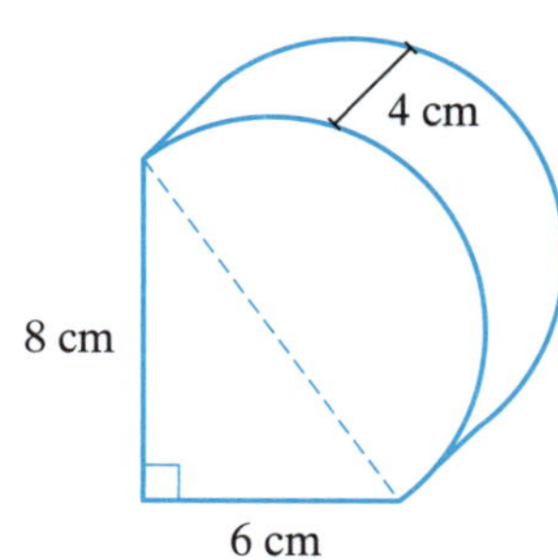

Find the volume, correct to 2 decimal places.

6 CHALLENGE

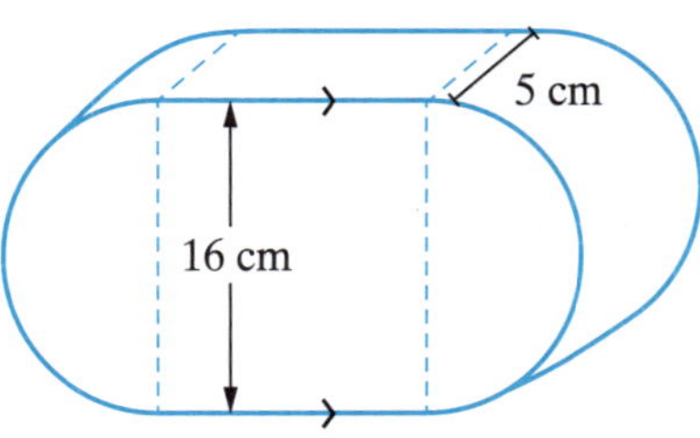

Find the volume in terms of π, if the surface area is $4(52\pi + 105)$ cm^2.

Answers page 150

KEY SKILL

38 Capacity

HINTS

- Capacity is the amount of liquid, or pourable substance, that a solid can hold.
- Make sure you know the **Volume of prisms and cylinders Hints** from Key Skills 35 to 37 on pages 88, 90 and 92.
- To convert between volume and capacity we use:
  - $1000\text{ cm}^3 = 1000\text{ mL} = 1\text{ L}$
  - $1\text{ m}^3 = 1000\text{ L} = 1\text{ kL}$.

  e.g. Find the capacity of a cube with side 8 cm.

  Volume $= 8^3 = 512$ $\quad \therefore 512\text{ cm}^3$

  Capacity $= 512$ mL

Examples

A cylindrical drum with diameter 12 cm and height 18 cm is half filled with water. Find the amount of water in the container in litres, correct to 3 significant figures.

Solution

Diameter 12 cm, radius 6 cm.

$$\begin{aligned} V &= \frac{1}{2} \times \pi r^2 h \\ &= \frac{1}{2} \times \pi \times 6^2 \times 18 \\ &= 1017.876\,02\ldots \\ &= 1017.9 \text{ (1 dec. pl.)} \end{aligned}$$

$\therefore$ volume is 1017.9 cm^3

Capacity is 1017.9 mL, or 1.02 L (3 sig. figs).

$\therefore$ contains 1.02 litres.

FOCUS on ...

1. The question: Asks you to find the amount of water in a drum.
2. The information: Gives you the diameter and the height of a cylinder and tells you that it is half full.
3. Your working: You need to divide the diameter by 2 to find the radius: $12 \div 2 = 6$.
   You need to find the volume by substituting $r = 6$ and $h = 18$ in $V = \frac{1}{2}\pi r^2 h$.
   $$\begin{aligned} V &= \frac{1}{2} \times \pi \times 6^2 \times 18 \\ &= 1017.876\,02\ldots \\ &= 1017.9 \text{ (1 dec. pl.)} \end{aligned}$$
   The volume is 1017.9 cm^3.
   To change to litres you divide by 1000.
   $1017.9 \div 1000 = 1.0179 = 1.02$ (3 sig. figs)
4. Your answer: Make sure you write the correct units: The drum contains 1.02 L.

An empty container is placed under a dripping tap. The container is in the shape of a cylinder with radius 20 cm and height 32 cm. The tap is dripping at the rate of 3 mL/5 seconds. How long will it take for the container to overflow?

Solution

$$\begin{aligned} V &= \pi r^2 h \\ &= \pi \times 20^2 \times 32 \\ &= 40\,212.386 \text{ (3 dec. pl.)} \end{aligned}$$

$\therefore$ volume is $40\,212.386\text{ cm}^3$

Capacity is 40 212.386 mL

$$\begin{aligned} \text{Rate} &= 3\text{ mL/5 s} = 0.6\text{ mL/s} \\ \text{No. of seconds} &= 40\,212.386 \div 0.6 \\ &= 67\,020.643\ldots \\ \text{No. of hours} &= 67\,020.643\ldots \div 60 \div 60 \\ &= 18.616\,845\,35\ldots \\ &= 18\text{ h }37\text{ min (nearest min.)} \end{aligned}$$

$\therefore$ it will take 18 h 37 min (or 18.62 h).

FOCUS on ...

1. The question: Asks you to find the time it takes to fill a cylinder.
2. The information: Gives you the radius and height of the cylinder and the rate of dripping water.
3. Your working: You need to find the volume by substituting $r = 20$ and $h = 32$ in $V = \pi r^2 h$.
   $$\begin{aligned} V &= \pi \times 20^2 \times 32 \\ &= 40\,212.385\,97\ldots \\ &= 40\,212.386 \text{ (3 dec. pl.)} \end{aligned}$$
   The volume is $40\,212.386\text{ cm}^3$.
   You write this in millilitres: 40 212.386 mL.
   Change the rate to per second: $3 \div 5 = 0.6$ mL/s.
   You divide to find the seconds.
   $40\,212.386 \div 0.6 = 67\,020.643\ldots$
   To change to minutes, divide this by 60, and then to change to hours, divide by 60 again:
   $67\,020.643\ldots \div 60 \div 60 = 18.616\,845\,37\ldots$
   Using the DMS key, this is 18 h 37 min (nearest min).
4. Your answer: Make sure you write the correct units: It will take 18 h 37 min, or 18.62 h.

Now try these!

1. A container in the shape of a cylinder has a diameter of 40 cm and a height of 85 cm. If the container is one-third filled, what is the amount of water in the container, to the nearest litre?

2. An empty tank in the shape of a rectangular prism is being filled with water at the rate of 1 L/5 seconds. The dimensions of the base are 1.4 m by 80 cm. How long would it take for the tank to be filled to a height of 1.1 m?

3. A cube of side 25 cm is filled with water. Ryan now empties the water into a cylinder of diameter 18 cm. What is the height of the water in the cylinder? Give your answer to the nearest centimetre.

4. Water from a 2-litre container of water is poured into a cylinder with a diameter of 12 cm and a height of 10 cm. When the cylinder is full, what percentage of water remains in the container? Give your answer to one decimal place.

5. A cylinder has a diameter of 24 cm and a height of 16 cm. The cylinder is three-quarters filled with water. This water is then poured into a cube. If the cube is half-filled, what are the dimensions of the cube, to the nearest centimetre?

6. CHALLENGE A cylinder with a diameter of 8 cm and a height of 12 cm is three-quarters filled with warm water. Jack drops ice-blocks measuring 4 cm by 3 cm by 2 cm into the container. How many ice-blocks can be placed into the container before water spills out?

Answers pages 150–151

KEY SKILL

39 Similar triangles

HINTS

- The matching sides of similar triangles are in proportion,
  e.g. If $\triangle ABC$ and $\triangle PQR$ are similar, write an equation linking matching sides.

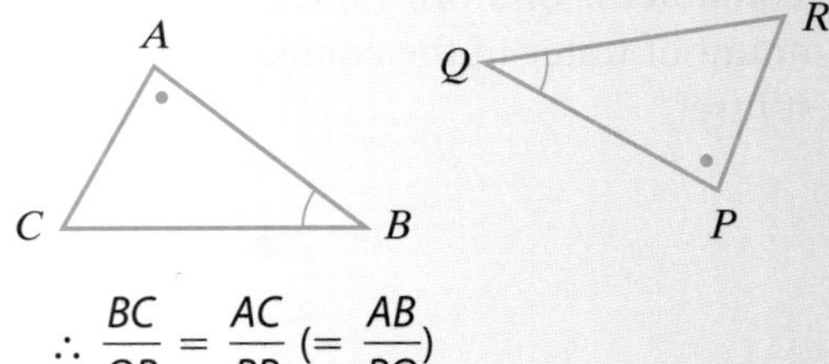

$$\therefore \frac{BC}{QR} = \frac{AC}{PR} \left(= \frac{AB}{PQ}\right)$$

Examples

Lara is 1.6 metres tall and one afternoon her friend measures Lara's shadow to be 2 metres. At that same time a nearby statue has a shadow that is 3.5 metres long. How tall is the statue?

Solution

The two triangles are similar.

$$\therefore \frac{x}{1.6} = \frac{3.5}{2}$$

$$x = \frac{1.6 \times 3.5}{2} = 2.8$$

$\therefore$ the statue is 2.8 metres tall.

FOCUS on ...

1. The question: Asks you to find the height of the statue.
2. The information: Gives you the relationship between Lara and her shadow, and the shadow of a statue.
3. Your working: Draw a diagram with 2 triangles to represent Lara and the statue.
   The two triangles are similar, so you need to set up an equation linking matching sides. Start with the unknown on the top on the LHS of the equation.
   $$\frac{x}{1.6} = \frac{3.5}{2}$$
   $$x = \frac{1.6 \times 3.5}{2}$$
   $$= 2.8$$
4. Your answer: Make sure you write the correct units:
   The statue is 2.8 metres tall.

In the diagram $\triangle ABD$ and $\triangle ACE$ are similar. If the area of $\triangle ABD$ is 54 cm^2, AB = 12 cm and $BD : CE$ = 3 : 2, what is the length of EC?

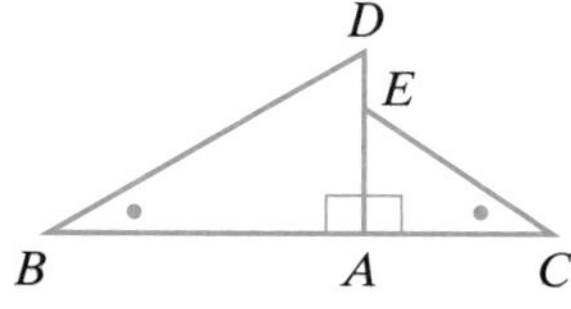

Solution

$$\frac{1}{2} \times 12 \times AD = 54$$
$$\therefore AD = 9 \text{ cm}$$

By Pythagoras: $BD^2 = 9^2 + 12^2$

$$= 225$$
$$BD = 15$$
$$3 \text{ parts} = 15$$
$$2 \text{ parts} = 15 \div 3 \times 2$$
$$= 10$$
$$\therefore EC = 10 \text{ cm}$$

FOCUS on ...

1. The question: Asks you to find the length of EC.
2. The information: Gives you the ratio of the sides of two similar triangles, the area of one triangle and the length of a side.
3. Your working: You find the side of the triangle using the given area and given side.
   $$\frac{1}{2} \times 12 \times AD = 54$$
   $$6 \times AD = 54$$
   $$\therefore AD = 9$$
   Apply Pythagoras' theorem.
   $$BD^2 = 9^2 + 12^2 = 225$$
   $$BD = 15$$
   All matching sides (e.g. BD and EC) are in the ratio 3:2.
   $$3 \text{ parts} = 15$$
   $$1 \text{ part} = 5$$
   $$2 \text{ parts} = 10$$
   $$\therefore EC = 10$$
4. Your answer: Make sure you write the correct units:
   The length of EC is 10 cm.

Now try these!

1 A wall with a height of 3.2 metres casts a shadow of 4.8 metres. At the same time a tree casts a shadow of 16 metres. What is the height of the tree, to 2 decimal places?

2

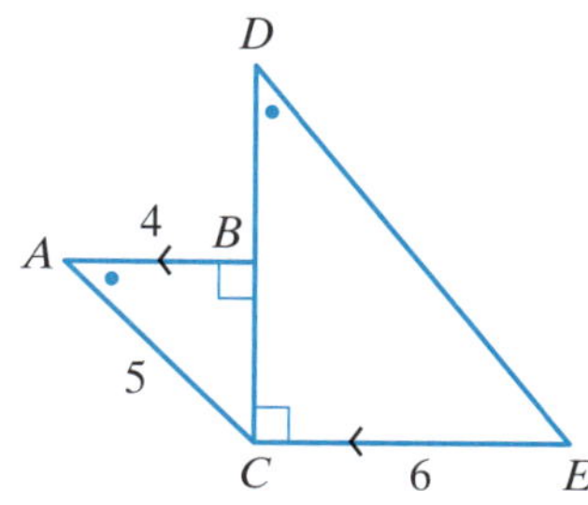

Find the length of DB.

3 A building of height 38 metres casts a shadow of 64 metres. A flagpole on the top of the building casts a shadow of 4 metres. How high is the top of the flagpole from the ground?

4

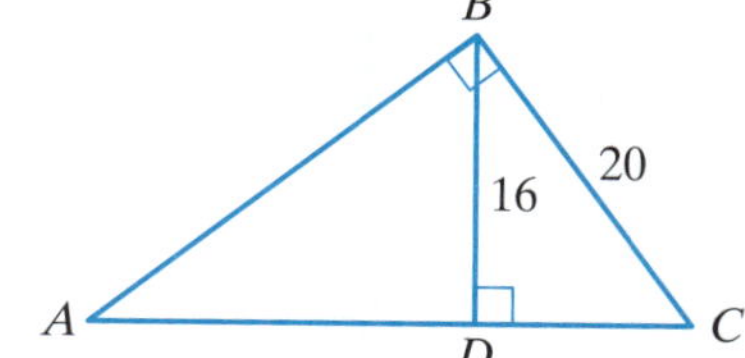

Find the length of AC.

5 Mia cuts a 4-mm diameter circular hole out of a piece of cardboard and holds the cardboard 45 cm from her eyes so that the moon just fits inside the circle. If the diameter of the moon is 3475 km, how far is the moon from Mia?

6 

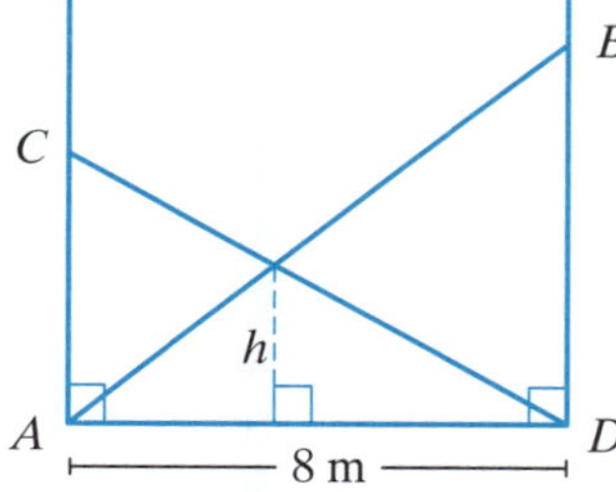

The diagram shows two ladders AB and DC, placed in a narrow alley. $AB = 17$ m and $DC = 10$ m and the alley is 8 metres wide. Find the height h, where the two ladders cross.

Answers pages 151–152

REVISION TEST 11 Level of difficulty—Average

1. A solid cylinder with diameter 12 cm and height 9 cm is 3D-printed using plastic with a cost of 0.522 cents/cm^3. Find the cost of the cylinder, to the nearest dollar.

2. Max has 16 shrubs in his garden that each require 1.1 litres of water each day from his automatic watering system. If he pays $1.18 per kilolitre for water, how much will it cost to water the shrubs each week?

3. 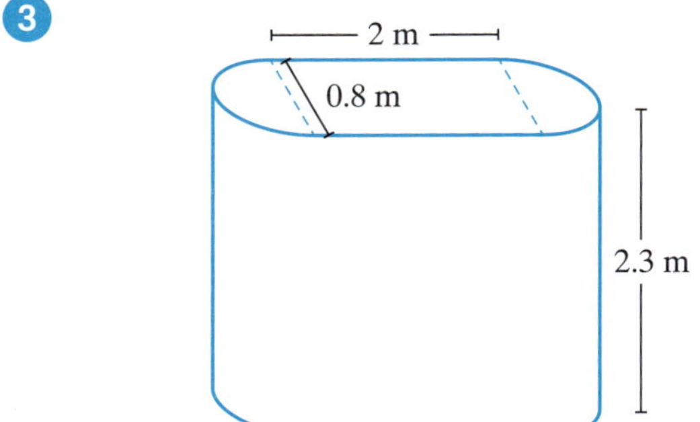

The diagram shows a water tank. Find the capacity of the tank, to the nearest litre.

4. The curved surface area of a cylinder is 60π cm^2 and the height is 5 cm. Find the exact volume of the cylinder.

5. 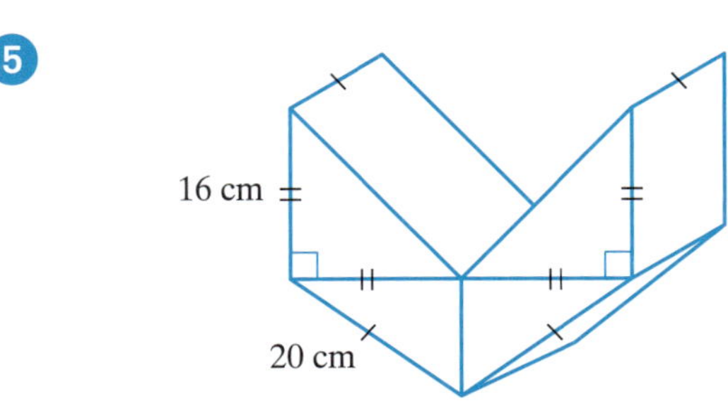

The diagram shows a fish tank. Find the capacity of the tank in litres.

6. 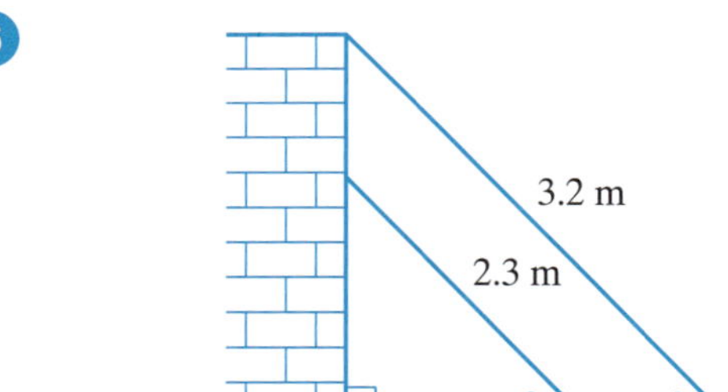

Two ladders lean against a wall so that they make the same angle with the ground. The 2.3 m ladder reaches 3 m up the wall. How much further does the other ladder reach, to 2 decimal places?

Answers page 152

1

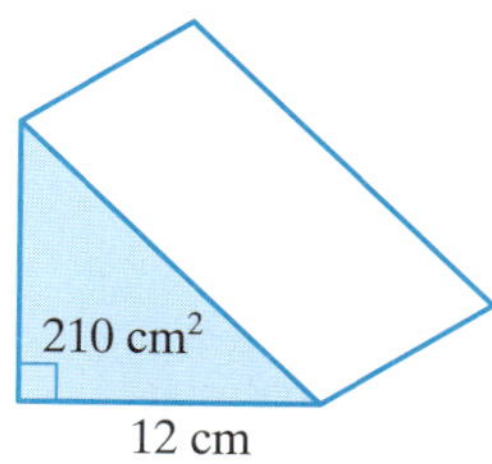

If the volume is 3360 cm^3, find the surface area.

2 An army's personnel carrier has a fuel tank in the shape of a rectangular prism measuring 95 cm by 50 cm by 40 cm. If the vehicle averages 14 litres/100 km, how far can the vehicle travel on a tank of petrol, to the nearest kilometre?

3

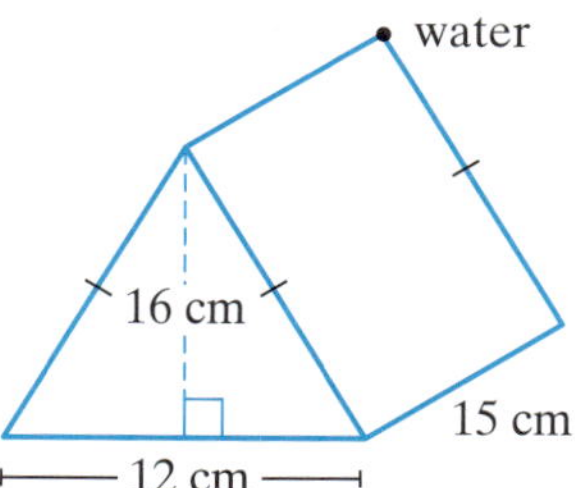

Water is poured into an empty triangular prism until the water is 6 cm deep. Find the amount of water in the prism in litres.

4

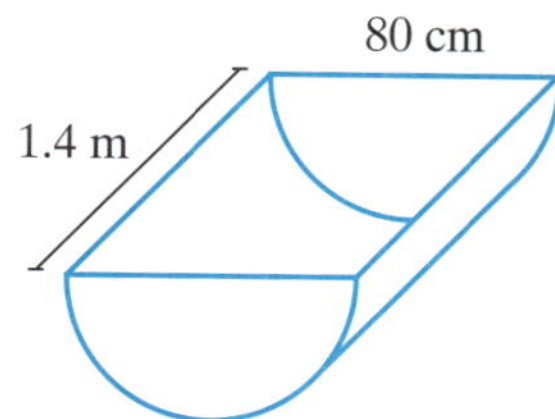

The diagram shows a bath. If water costs \$2.23/kL, find the cost of filling the bath 50 times.

5 A cylinder has a total surface area of 66π cm^2. If the diameter is 8 cm, what is the capacity, to the nearest millilitre?

6

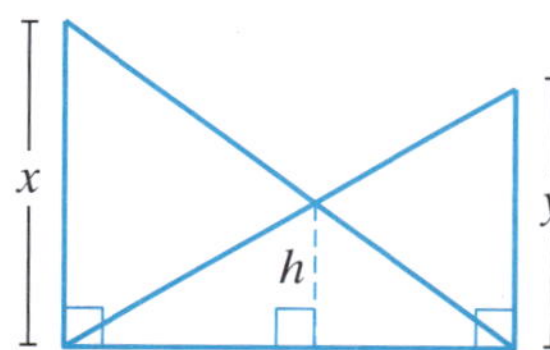

Two ladders are placed on opposite diagonals in a narrow alley such that one ladder reaches x m up one wall, the other ladder reaches y m up the opposite wall and they intersect h m above the ground.

Prove that $\frac{1}{x} + \frac{1}{y} = \frac{1}{h}$.

Answers pages 152–153

KEY SKILL

40 Trigonometry: Finding sides

HINTS

- In a right triangle, the ratio of sides and angles are linked:

opposite hypotenuse θ adjacent

$$\sin\theta = \frac{\text{opposite}}{\text{hypotenuse}} \quad \cos\theta = \frac{\text{adjacent}}{\text{hypotenuse}} \quad \tan\theta = \frac{\text{opposite}}{\text{adjacent}}$$

e.g. Evaluate $25 \times \tan 18° 23'$, to 3 decimal places.

$= 8.308\,316\,943\ldots$

$= 8.308$ (3 dec. pl.)

Examples

A ladder makes an angle of 72° with the ground and the base of the ladder is 0.8 metres from a wall. Find the length of the ladder, correct to 2 decimal places.

Solution

Let the length be x.

$$\frac{0.8}{x} = \cos 72°$$

$$x = \frac{0.8}{\cos 72°}$$

$= 2.588\,543\,82\ldots$

$= 2.59$ (2 dec. pl.)

∴ the ladder is 2.59 m long.

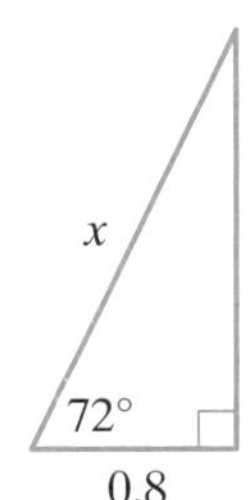

FOCUS on ...

1. The question: Asks you to find the length of the ladder.
2. The information: Gives you the distance a ladder is from a wall and the angle it makes with the ground.
3. Your working: Draw a triangle to represent the information. Let the length of the ladder = x.
   From the given angle, the sides involved are adjacent and hypotenuse. This means you will use cos:

$$\frac{0.8}{x} = \cos 72°$$

$$x \times \frac{0.8}{x} = x \times \cos 72°$$

$$0.8 = x \times \cos 72°$$

$$x = \frac{0.8}{\cos 72°}$$

$= 2.588\,543\,82\ldots$

$= 2.59$ (2 dec. pl.)

4. Your answer: Make sure you write the correct units: The ladder is 2.59 m long.

The length of Sailor Sam's anchor rope is 45 metres. It is attached to the bottom of the lake but after the boat has drifted in the current, the angle the rope makes with the surface of the water is 55° 30′. At what depth is Sam's anchor, correct to 3 significant figures?

Solution

Let the depth be x.

$$\frac{x}{45} = \sin 55°30'$$

$x = 45 \times \sin 55°30'$

$= 37.085\,678\,49\ldots$

$= 37.1$ (3 sig. figs)

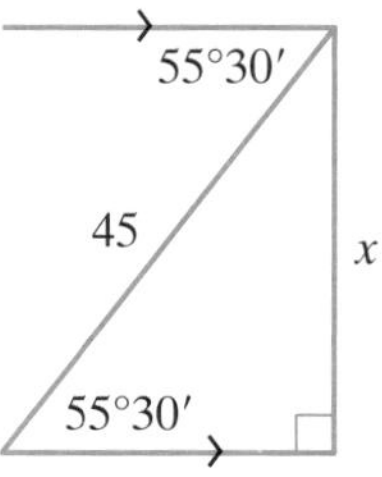

∴ the anchor is at a depth of 37.1 metres.

FOCUS on ...

1. The question: Asks you to find the depth of the anchor.
2. The information: Gives you the length of the rope and the angle the rope makes with the surface.
3. Your working: Draw a triangle to represent the information. Let the length of the depth = x.
   Use alternate angles and parallel lines to find angle inside triangle. From the given angle, the sides involved are opposite and hypotenuse. This means you will use sin:

$$\frac{x}{45} = \sin 55°30'$$

$x = 45 \times \sin 55°30'$

$= 37.085\,678\,49\ldots$

$= 37.1$ (3 sig. figs)

4. Your answer: Make sure you write the correct units: The anchor is 37.1 m underwater.

Now try these!

1. A kite is anchored to the ground and is flying at a height of 32 metres. If the string makes an angle of 42° with the ground, how long is the string? Answer to 3 significant figures.

2. 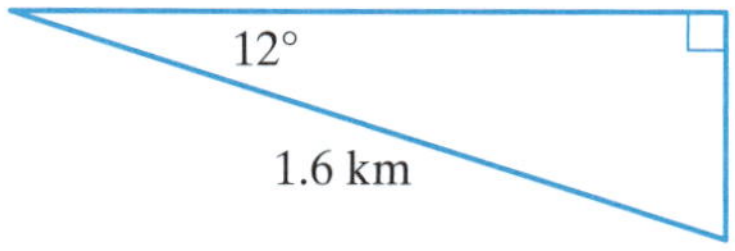 12°

   1.6 km

   A mine shaft is dug at an angle of 12°. If the mine shaft is 1.6 kilometres long, how far below the surface is the end of the shaft? Give your answer to the nearest metre.

3. For an 18-metre pole to be stable, an engineer decided that three equally-spaced guide wires from the top of the pole to the ground should be used. If the wires make a 75° angle with the ground, find the length of wire required, to the nearest metre.

4. 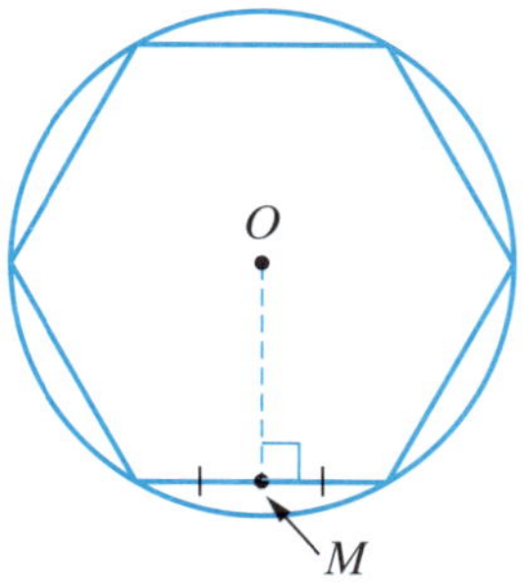 O

   M

   A regular hexagon is drawn inside a circle with radius 8 cm. M is the midpoint of one of the sides. Find the length of the apothem OM, correct to 2 decimal places.

5. A plane takes off at an angle of 11° to the horizontal. If the plane maintains a speed of 270 km/h, what horizontal distance has it travelled after flying for 2 minutes?

6. CHALLENGE A disability ramp is to be built using an angle of '1 in 14'. The ramp needs to rise 0.8 metres and is to be 1.2 metres wide. What is the cost of covering the ramp with non-slip material at $3.40/m^2? (Answer to the nearest dollar.)

Answers page 153

KEY SKILL

41 Trigonometry: Finding angles

HINTS

- In a right triangle, the ratio of sides and angles are linked:

$\sin\theta = \dfrac{\text{opposite}}{\text{hypotenuse}}$ $\cos\theta = \dfrac{\text{adjacent}}{\text{hypotenuse}}$ $\tan\theta = \dfrac{\text{opposite}}{\text{adjacent}}$

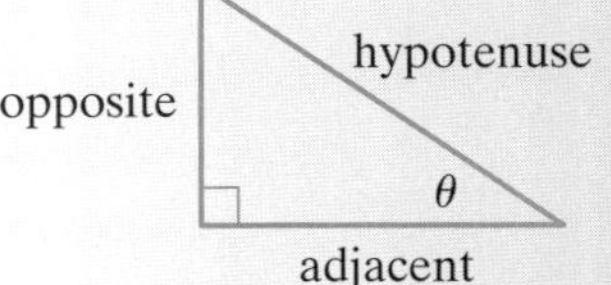

e.g. If θ is acute, find θ, if $\tan\theta = 1.3216$.
Leave your answer to the nearest minute.

$$\tan\theta = 1.3216$$
$$\theta = 52.886\,715\,56\ldots$$
$$= 52°53' \quad \text{(nearest minute)}$$

Examples

One end of a 6.4 metre see-saw is sitting on the ground and the other end is 1.8 metres high. What angle does the see-saw make with the ground? Answer to the nearest minute.

Solution

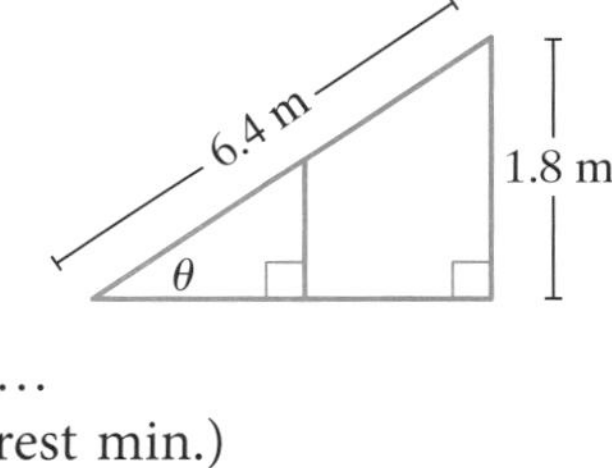

Let the angle be θ.

$$\sin\theta = \frac{1.8}{6.4}$$
$$\theta = 16.334\,8227\ldots$$
$$= 16°20' \text{ (nearest min.)}$$

$\therefore$ the see-saw makes an angle of $16°20'$ with the ground.

FOCUS on...

1. The question: Asks you to find the angle between the see-saw and the ground.
2. The information: Gives you the length of a see-saw and the height above the ground of one end.
3. Your working: Draw a triangle to represent the information. Let the unknown angle = θ.
   From the unknown angle, the known sides are the opposite and hypotenuse. This means you will use sin:
   $$\sin\theta = \frac{1.8}{6.4}$$
   $$\theta = 16.334\,8227\ldots$$
   $$= 16°20'$$
4. Your answer: Make sure you write the correct units:
   The see-saw makes an angle of 16° 20′ with the ground.

An isosceles triangle has sides of 6 cm, 6 cm and 8 cm. Find the size of each angle, to the nearest minute.

Solution

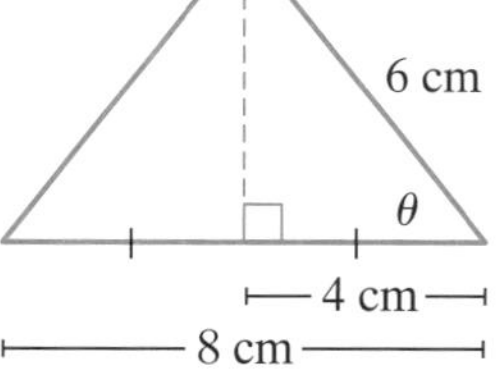

Let the angle be θ.

$$\cos\theta = \frac{4}{6}$$
$$\theta = 48.189\,6851\ldots$$
$$= 48°11' \text{ (nearest min.)}$$
$$180° - 2 \times 48°11' = 83°38'$$

$\therefore$ the angles are $48°11'$, $48°11'$ and $83°38'$.

FOCUS on...

1. The question: Asks you to find the angles in an isosceles triangle.
2. The information: Gives you the lengths of the sides.
3. Your working: Draw a triangle to represent the information. Let the unknown angle = θ.
   Draw a line from the vertex to the middle of the base. In an isosceles triangle, this line forms a right angle. From the unknown angle, the known sides are the adjacent and hypotenuse. This means you will use cos:
   $$\cos\theta = \frac{4}{6}$$
   $$\theta = 48.189\,6851\ldots$$
   $$= 48°11'$$
   The other base angle is also 48°11′.
   To find the third angle you subtract the total of the two angles from 180°.
   $180° - 2 \times 48°11' = 83°38'$
4. Your answer: Make sure you write the correct units:
   The angles are 48°11′, 48°11′ and 83°38′.

Now try these!

1. A 3.2-metre ladder rests against a vertical wall so that the distance from the base of the ladder to the wall is 0.7 metres. What is the size of the angle that the top of the ladder makes with the wall? Leave your answer to the nearest minute.

2. Find the size of each angle of a triangle with the dimensions 12 cm, 12 cm and 16 cm. Give your answer to nearest degree.

3. Two boats leave a dock at the same time. Boat A goes south for 3 km and stops. Boat B travels west for 2 km and stops and turns towards Boat A. What angle must Boat A turn to face and go directly to Boat B? Give your answer to the nearest minute.

4. 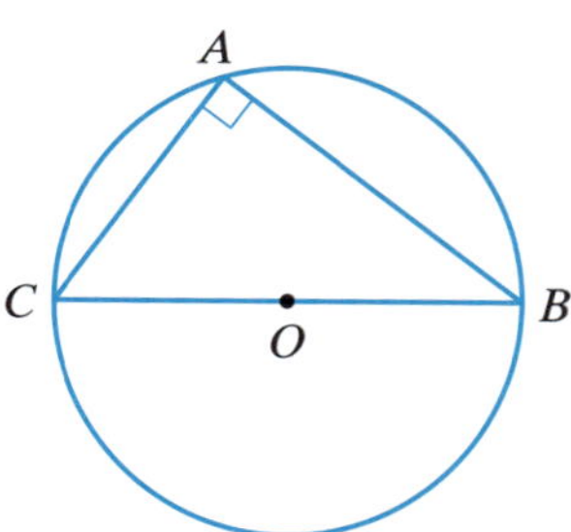

A circle has area 16π cm^2 and chord $AB = 6$ cm. Find the size of $\angle ACB$, to the nearest minute.

5. An aeroplane is approaching an airport at a height of 1100 metres. If the aeroplane is currently 7 kilometres horizontally from its landing point, what is the angle of approach, to the nearest minute?

6. CHALLENGE

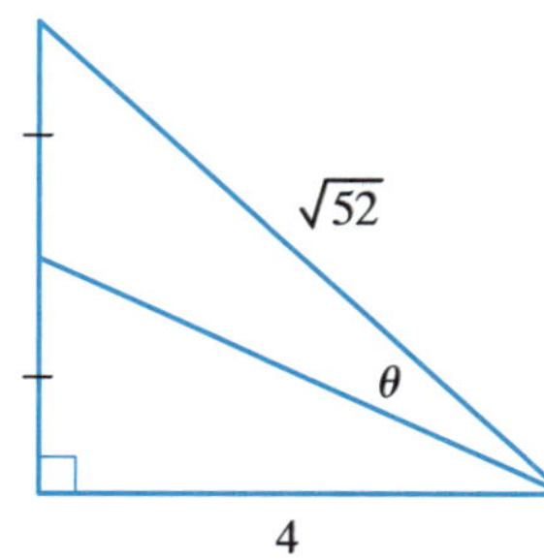

Find the size of θ, to the nearest minute.

Answers page 154

KEY SKILL

42 Trigonometry: Angle of elevation

HINTS

- Make sure you know the **Trigonometry: Finding sides and angles Hints** from Key Skills 40 and 41 on pages 100 and 102.
- The angle of elevation is the angle above the horizontal and the angle of depression is the angle below the horizontal.

  The two angles are equal (alternate angles, parallel lines).

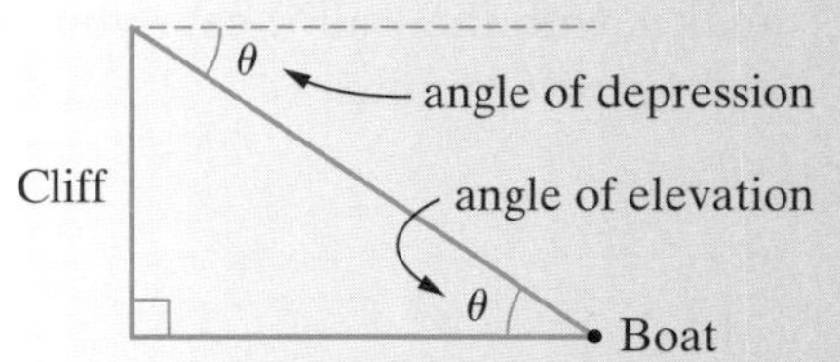

Examples

Benjamin is walking to his 148-metre tall office building. He stops and measures the angle of elevation of the top of the building to be 63°. How much further does Benjamin need to walk before he has reached the building, to the nearest metre?

Solution

Let the distance be x.

$$\frac{148}{x} = \tan 63°$$

$$x = \frac{148}{\tan 63°}$$

$$= 75.409\,766\,53\ldots$$

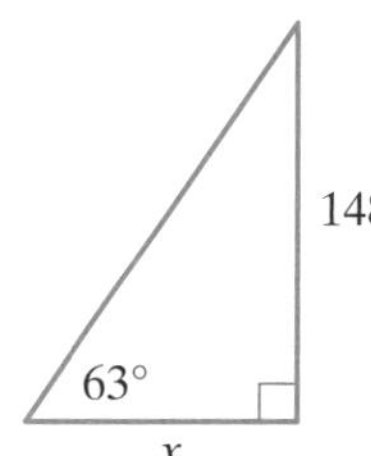

$$= 75 \text{ (nearest whole)}$$

$\therefore$ Benjamin has to walk another 75 metres.

FOCUS on ...

1. The question: Asks you to find the distance.
2. The information: Gives you the height of a building and the angle of elevation of the top of the building.
3. Your working: Draw a triangle to represent the information. The angle of elevation is the angle formed between the horizontal and the line of sight of the top of the building. Let the length of the unknown side = x. From the given angle, the sides involved are opposite and adjacent. This means you will use tan:

   $$\frac{148}{x} = \tan 63°$$

   $$x = \frac{148}{\tan 63°}$$

   $$= 75.409\,766\,53\ldots$$

   $$= 75 \text{ (nearest whole)}$$
4. Your answer: Make sure you write the correct units:

   Benjamin walks 75 metres.

A tree 12 metres tall casts a shadow 16.5 metres long. What is the angle of elevation of the sun, to the nearest minute?

Solution

Let the angle be θ.

$$\tan \theta = \frac{12}{16.5}$$

$$\theta = 36.027\,373\,39\ldots$$

$$= 36°2' \text{ (nearest minute)}$$

12

θ

16.5

$\therefore$ the angle of elevation of the sun is $36°2'$.

FOCUS on ...

1. The question: Asks you to find the angle of elevation of the sun.
2. The information: Gives you the height of a tree and the length of its shadow.
3. Your working: Draw a triangle to represent the information. The angle of elevation is the angle formed between the horizontal at the end of the shadow and the line of sight of the top of the tree. Let the unknown angle = θ.

   From the unknown angle, the known sides are the adjacent and opposite. This means you will use tan:

   $$\tan \theta = \frac{12}{16.5}$$

   $$\theta = 36.027\,373\,39\ldots$$

   $$= 36°2' \text{ (nearest minute)}$$
4. Your answer: Make sure you write the correct units:

   The angle of elevation of the sun is $36°2'$.

Now try these!

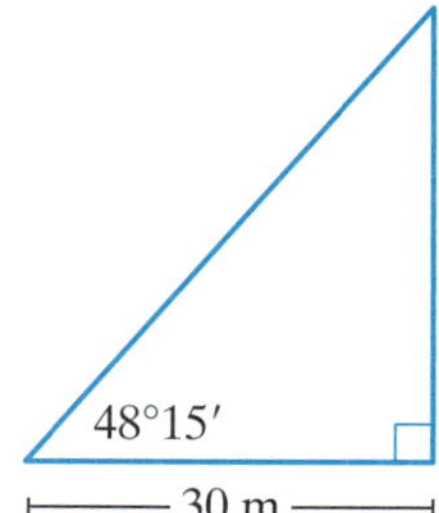

Jocelyn wants to find the height of a tree. She walks 30 metres from the base of the tree, turns and measures the angle of elevation of the top of the tree as 48°15′. How tall is the tree, to the nearest centimetre?

2 A 10-metre high flagpole casts a 9-metre long shadow. What is the angle of elevation of the sun, to the nearest minute?

3 Two buildings with flat roofs are 25 metres apart. From the top of the shorter building (which is 32 metres high) the angle of elevation to the edge of the taller building is 38°20′. How high is the taller building, to the nearest metre?

4 Jannah is observing a vertical rocket launch. When she first sees the rocket it has an angle of elevation of 14°. When she stops looking at the rocket it has an angle of elevation of 76°. If she is 2 km from the launch site how far did the rocket travel while Jannah was watching it? Leave your answer to the nearest metre.

5 From a point 180 metres away, the angle of elevation of the top of a building is 50°. What would be the angle of elevation of a point half-way down the building? Give your answer to the nearest degree.

6 **CHALLENGE** An aeroplane is cruising at an altitude of 10 000 metres. It is flying in a straight line away from Anne who is standing on the ground. She notices that in one minute, the aeroplane flies from being directly overhead to having an angle of elevation of 25°. What is the speed of the aeroplane, to the nearest kilometre per hour?

Answers pages 154–155

KEY SKILL

43 Trigonometry: Angle of depression

HINTS

- Make sure you know the **Trigonometry: Finding Sides and angles Hints** from Key Skills 40 and 41 on pages 100 and 102 and the **Trigonometry: Angle of elevation Hints** from Key Skill 42 on page 104.

Reminder!

- Read the question carefully to identify what needs to be found.
- Re-read the question when you've finished your answer to make sure that the question has been answered and your solution makes sense.

Examples

An aeroplane is flying at a height of 3000 metres. The distance along the ground from the aeroplane to an airport is 16 km. What is the angle of depression from the aeroplane to the airport? Give your answer to the nearest minute.

Solution

Let the angle be θ.

$$\tan\theta = \frac{3}{16}$$

$$\theta = 10.619\,655\,28\ldots$$

$$= 10°37' \text{ (nearest minute)}$$

$\therefore$ the angle of depression is $10°37'$.

FOCUS on ...

1. The question: Asks you to find the angle of depression.
2. The information: Gives you the height of a plane and the distance to an airport.
3. Your working: Draw a triangle to represent the information. The angle of depression is the angle formed between the horizontal and the line of sight of the airport. Using alternate angles and parallel lines, the angle is equal to an angle inside the triangle. Let the unknown angle = θ.
   From the unknown angle, the known sides are the opposite and adjacent. This means you will use tan:
   $$\tan\theta = \frac{3}{16}$$
   $$\theta = 10.619\,655\,28\ldots$$
   $$= 10°37' \text{ (nearest minute)}$$
4. Your answer: Make sure you write the correct units: The angle of depression is $10°37'$.

A kookaburra is flying at a height of 12 metres and spots a 1.8-metre fence on which to perch. If the top of the fence is at an angle of depression of 36°15′, how far must the bird fly before it can land? Give your answer to the nearest metre.

Solution

$12 - 1.8 = 10.2$

Let the distance be x.

$$\frac{10.2}{x} = \sin 36°15'$$

$$x = \frac{10.2}{\sin 36°15'}$$

$$= 17.249\,845\,37\ldots$$

$$= 17 \text{ (nearest whole)}$$

$\therefore$ the kookaburra has to fly 17 metres.

FOCUS on ...

1. The question: Asks you to find the distance flown by the bird.
2. The information: Gives you the height of the bird, the height of a fence and the angle of depression from the bird to the top of the fence.
3. Your working: Draw a triangle to represent the information. The angle of depression is the angle formed between the horizontal and the line of sight of the fence. Using alternate angles and parallel lines, the angle is equal to an angle inside the triangle. Let the length of the unknown side = x. You subtract to find the height above the fence: 12 − 1.8 = 10.2.
   From the given angle, the sides are the opposite and hypotenuse. This means you will use sin:
   $$\frac{10.2}{x} = \sin 36°15'$$
   $$x = \frac{10.2}{\sin 36°15'} = 17 \text{ (nearest whole)}$$
4. Your answer: Make sure you write the correct units: The bird flies 17 metres.

Now try these!

1. Ian is standing on top of a 45-metre-high cliff and he sees a swimmer, 25 metres from the base of the cliff. What is the angle of depression to the swimmer, to the nearest degree?

2. The pilot of a helicopter measures the angle of depression of the landing spot to be $16°10'$. If the altitude of the helicopter is 1800 metres, what is the horizontal distance to the landing spot? Give your answer to the nearest metre.

3. Jordie is standing on the ground and notices his wife Lyndy in a hot-air balloon directly above a point 450 metres away. Lyndy is told by the balloon's pilot that she is flying at an altitude of 360 metres. What is the angle of depression of Jordie from Lyndy, correct to the nearest minute?

4. A phone tower is 30 metres high and is built on a mountain 460 metres above sea level. What is the angle of depression from the top of the tower to a phone user who is 8 kilometres away and is 120 metres above sea level? Give your answer to the nearest minute.

5. The base of a lighthouse is at the top of a cliff 125 metres above sea level. The angle of depression from the base of the lighthouse to a boat is 30°. The boat then travels towards the cliff and stops 110 metres from the base of the cliff. What is the angle of depression of the new location of the boat from the base of the lighthouse, to the nearest minute?

6. CHALLENGE From a hillside lookout the angle of depression to the far side of a dam is 42° and the angle of depression of the near side of the dam is 61°. If the lookout is 120 metres above the dam, what is the width of the dam, to the nearest metre?

Answers page 155

KEY SKILL

44 Trigonometry: Bearings A

HINTS

- A bearing to a point is an angle, expressed in 3 digits, measured in a clockwise direction from north.

  e.g. What is the bearing of A from B and the bearing of B from A?

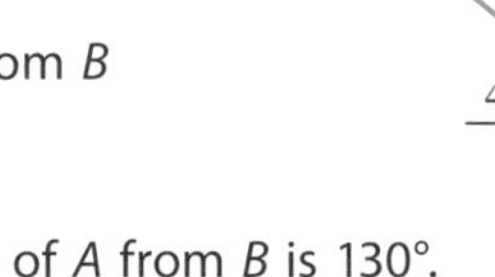

As $90 + 40 = 130$, the bearing of A from B is 130°.
As $270 + 40 = 310$, the bearing of B from A is 310°.

Reminder!

- Read the question carefully to identify what needs to be found.
- Re-read the question when you've finished your answer to make sure that the question has been answered and your solution makes sense.

Examples

A ship leaves Port Hedland and travels 380 km on a bearing of 335°. How far north of Port Hedland is the ship, to the nearest kilometre?

Solution

$335 - 270 = 65$

Let the distance be x.

$$\frac{x}{380} = \sin 65°$$
$$x = 380 \times \sin 65°$$
$$= 344.396\,9591\ldots$$
$$= 344 \text{ (nearest whole)}$$

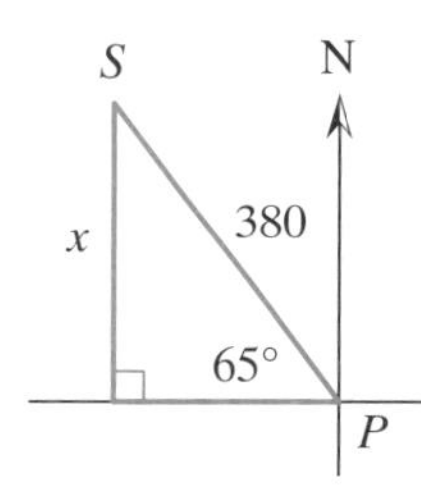

$\therefore$ the ship is 344 km north of Port Hedland.

FOCUS on...

1. The question: Asks you to find the distance the ship is north of Port Hedland.
2. The information: Gives you the distance and bearing the boat has travelled.
3. Your working: Draw a triangle to represent the information. As the bearing is 335°, you need to subtract 270° to find the angle inside the triangle. Let the distance north of Port Hedland = x.

   From the given angle, the sides are the opposite and hypotenuse. This means you will use sin:

   $$\frac{x}{380} = \sin 65°$$
   $$x = 380 \times \sin 65°$$
   $$= 344.396\,9591\ldots$$
   $$= 344 \text{ (nearest whole)}$$
4. Your answer: Make sure you write the correct units: The ship is 344 km north of Port Hedland.

A plane takes off from an airstrip and flies 160 km east and then turns and flies 130 km south. What is the bearing of the plane from the airstrip? Give your answer to the nearest degree.

Solution

Let the angle be θ.

$$\tan \theta = \frac{130}{160}$$
$$\theta = 39.093\,858\,89\ldots$$
$$= 39 \text{ (nearest whole)}$$

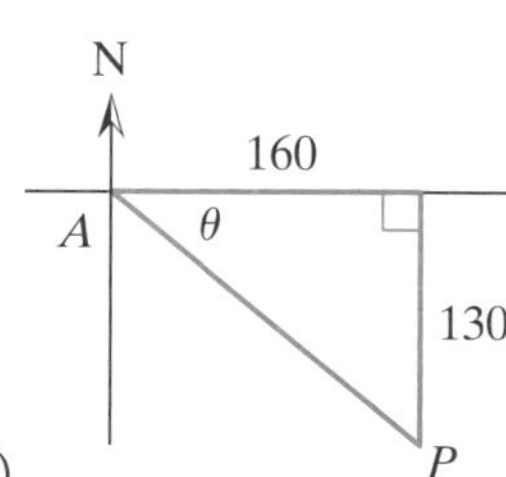

$90 + 39 = 129$

$\therefore$ the bearing is 129°.

FOCUS on...

1. The question: Asks you to find the bearing of the plane from the airstrip.
2. The information: Gives you two distances the plane has flown.
3. Your working: Draw a triangle to represent the information. Let the unknown angle = θ.

   From the unknown angle, the given sides are the opposite and adjacent. This means you will use tan:

   $$\tan \theta = \frac{130}{160}$$
   $$\theta = 39.093\,858\,89\ldots$$
   $$= 39 \text{ (nearest whole)}$$

   The plane has flown 39° south of east. This means you need to add: $90 + 39 = 129$.
4. Your answer: Make sure you write the correct units: The plane is flying on a bearing of 129°.

Now try these!

1. A ship leaves Fremantle and sails on a bearing of 348° for 860 km. How far west of Fremantle is the ship? Give your answer to the nearest kilometre.

2. At dawn, Cameron left his campsite and walked 6 km south and then 3 km west looking for water. What is his bearing from his overnight campsite? Give your answer to the nearest degree.

3. Jans left Packerere to drive on a bearing of 228° on a straight road towards Lemontree at an average speed of 72 km/h. After 2 hours, how far south has he travelled? Give your answer to 2 decimal places.

4. Alberka is 26 km south of Berridale and Carlsberg is 23 km east of Berridale. What is the bearing of Alberka from Carlsberg? Give your answer to the nearest degree.

5. In a playground, Holly left point P, walked 8 metres south to Q, 12 metres east to R, 20 metres north to S and 7 metres west to T. Find her distance from point P and also the bearing of T from P.

6. CHALLENGE Two cars left an intersection at 7:30 am. Car A headed south at a constant speed of 80 km/h. Car B travelled east and both cars stopped at 10 am. If the bearing of Car A from Car B is 230°, how much faster was Car B travelling than Car A, to the nearest kilometre per hour?

Answers pages 155–156

KEY SKILL

45 Trigonometry: Bearings B

HINTS

- Make sure you know the **Trigonometry: Bearings Hints** from Key Skill 44 on page 108.

Examples

A sailboat travels for half an hour at an average speed of 6 km/h on a bearing of 52° and then travels for 40 minutes at an average speed of 9 km/h on a bearing of 142°. How far is the sailboat from its starting point?

Solution

$180 - 142 = 38$

$52 + 38 = 90$

Using $D = ST$:

$6 \times \frac{1}{2} = 3$

$9 \times \frac{2}{3} = 6$

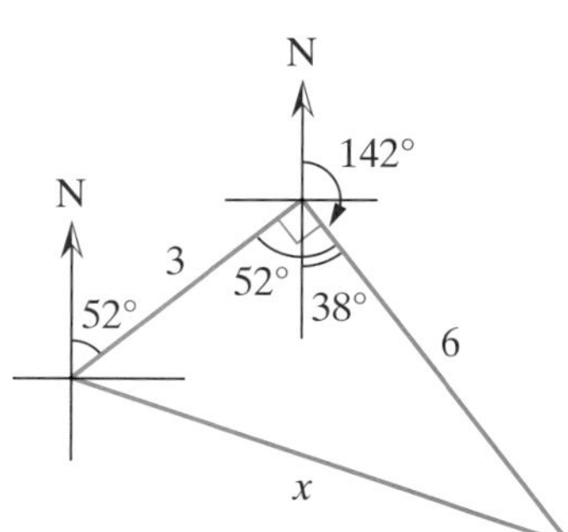

Let the distance be x.

$$x^2 = 3^2 + 6^2 = 9 + 36 = 45$$

$x = \sqrt{45}$, or 6.71 (2 dec. pl.)

∴ the sailboat is 6.71 km from its starting point.

FOCUS on ...

1. The question: Asks you to find the distance the boat has sailed.
2. The information: Gives you the speed, time and bearing of a boat before and after it changes directions.
3. Your working: Draw a triangle to represent the information using 2 compass roses. You need to calculate the angles on the diagram.
   As $180 - 142 = 38$, and $52 + 38 = 90$, the boat turned 90° when it changed directions.
   Using $D = ST$, the two sides can be calculated.
   $6 \times \frac{1}{2} = 3$, and as 40 min $= \frac{2}{3}$ h, then $9 \times \frac{2}{3} = 6$.
   Let the distance $= x$. Using Pythagoras:
   $$x^2 = 3^2 + 6^2 = 45$$
   $$x = \sqrt{45} = 6.708\,203\,932\ldots = 6.71 \text{ (2 dec. pl.)}$$
4. Your answer: Make sure you write the correct units: The boat has sailed 6.71 km.

A plane flies on a bearing of 150° for 8 km and then turns and flies on a bearing of 240° for 12 km. What is its bearing from its starting point, to the nearest degree?

Solution

$150 - 90 = 60$

$270 - 240 = 30$

$60 + 30 = 90$

Let the angle be θ.

$$\tan\theta = \frac{12}{8}$$

$$\theta = 56.309\,932\,47\ldots = 56$$

$150 + 56 = 206$

∴ the bearing is 206°.

FOCUS on ...

1. The question: Asks you to find the bearing of the plane from its starting point.
2. The information: Gives you the distance and bearing of a plane before and after it changes directions.
3. Your working: Draw a triangle to represent the information using 2 compass roses. You need to calculate the angles on the diagram:
   As $150 - 90 = 60$, and $270 - 240 = 30$, the plane turned 90° when it changed directions.
   Let the unknown angle $= \theta$.
   From the unknown angle, the given sides are the opposite and adjacent. This means you will use tan:
   $$\tan\theta = \frac{12}{8}$$
   $$\theta = 56.309\,932\,47\ldots = 56 \text{ (nearest whole)}$$
   Bearing $= 150 + 56 = 206$.
4. Your answer: Make sure you write the correct units: The bearing of the plane from its starting point is 206°.

Now try these!

1. Lee walks for 40 minutes at 6 km/h on a bearing of 120° and then stops and then jogs at 10 km/h for 30 minutes on a bearing of 210°. How far is Lee from his starting position, to the nearest metre?

2. A boat leaves a buoy and travels at 12 km/h on a bearing of 310° for 2 hours and then changes to travel on a bearing of 040° for 1 hour at 16 km/h. What is the bearing of the buoy from the boat?

3. A plane flies for 2 hours 10 minutes at 360 km/h on a bearing of 137°. It then turns and flies at the same speed for 1 hour 40 minutes on a bearing of 227°. How far is the plane from its starting point, to the nearest kilometre?

4. A boat sails the first leg of a course on a bearing of 125° and then turns and sails the second leg on a bearing of 035°. Find the bearing of the boat from the starting point if the boat sails 1800 metres on each leg.

5. The bearing of Sandham from Tregal is 125°. The bearing of Potomac from Sandham is 215°. The bearing of Tregal from Potomac is 028° and the distance from Tregal to Potomac is 38 km. Find the distance from Potomac to Sandham, to the nearest kilometre.

6. CHALLENGE B is 80 km from A, and the bearing of B from A is 110°. C is located on a bearing of 050° from A and 020° from B. Regan left A at 11:30 am and travelled to C by first passing through B. If he averaged 90 km/h for the entire trip, what time did he arrive in C?

Answers pages 156–157

KEY SKILL

46 Trigonometry: Two triangles A

HINTS

- Make sure you know the **Trigonometry Hints** from Key Skills 40 to 45 on pages 100, 102, 104, 106, 108 and 110.

Examples

Cassie is 15 metres from a tower and the angle of elevation of the top of the tower is 68°. David is on the other side of the tower and the angle of elevation of the top of the tower is 52°. How far is David from the base of the tower, to the nearest metre?

Solution

Let sides be h and x.

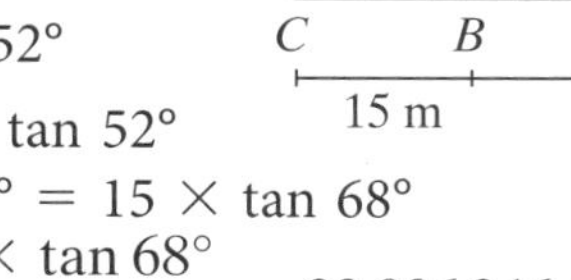

$$\frac{h}{15} = \tan 68°$$

$$h = 15 \times \tan 68°$$

$$\frac{h}{x} = \tan 52°$$

$$h = x \times \tan 52°$$

$$x \times \tan 52° = 15 \times \tan 68°$$

$$x = \frac{15 \times \tan 68°}{\tan 52°} = 29.006\,246\,75\ldots$$

$$= 29 \text{ (nearest whole)}$$

$\therefore$ David is 29 metres from the tower.

FOCUS on ...

1. The question: Asks you to find the distance David is from the base of the tower.
2. The information: Gives you the distance Cassie is from the tower and the angles of elevation of the top of the tower from Cassie and David.
3. Your working: Draw a triangle to represent the information. Let the height of the tower be h and the distance David is from the tower be x.
   From $\triangle ACB$, find an expression for h by using tan:
   $$\frac{h}{15} = \tan 68° \quad \therefore h = 15 \times \tan 68° \ldots ①$$
   From $\triangle ADB$, find an expression for h by using tan:
   $$\frac{h}{x} = \tan 52° \quad \therefore h = x \times \tan 52° \ldots ②$$
   Let ② = ①: $x \times \tan 52° = 15 \times \tan 68°$
   $$x = \frac{15 \times \tan 68°}{\tan 52°}$$
   $$= 29.006\,246\,75\ldots = 29 \text{ (nearest whole)}$$
4. Your answer: Make sure you write the correct units: David is 29 metres from the base of the tower.

Megan is at Town A which is 26 km from Town B. The bearing of B from A is 062°. She then travels east to Town C. If the bearing of Town B from Town C is 313°, how far is Town B from Town C, to the nearest kilometre?

Solution

Let sides be x and y.

$$\frac{x}{26} = \sin 28°$$

$$x = 26 \times \sin 28°$$

$$\frac{x}{y} = \sin 43°$$

$$y = \frac{x}{\sin 43°}$$

$$= \frac{26 \times \sin 28°}{\sin 43°} = 17.897\,7859\ldots$$

$= 18$ (nearest whole) $\therefore$ 18 km

FOCUS on ...

1. The question: Asks you to find the distance between two towns.
2. The information: Gives you the distances and bearings of three towns.
3. Your working: Draw a triangle to represent the information using 2 compass roses. You need to calculate the angles in the triangles: $90 - 62 = 28$ and $313 - 270 = 43$.
   To find the distance BC (called y), you first need to find the common side of the two triangles (called x).
   From the given angle (28°), the sides are the opposite and hypotenuse. This means you will use sin:
   $$\frac{x}{26} = \sin 28°$$
   $$x = 26 \times \sin 28°$$
   In the second triangle, from the given angle (43°), the sides are the opposite and hypotenuse.
   $$y = \frac{x}{\sin 43°}$$
   $$= \frac{26 \times \sin 28°}{\sin 43°} = 18 \text{ (nearest whole)}$$
4. Your answer: Make sure you write the correct units: Town B and Town C are 18 km apart.

Now try these!

1. Nick is 60 metres from a pole and the angle of elevation of the top of the pole is 14°. William is on the other side of the pole and the angle of elevation of the top of the pole from his position is 24°. How far is William from the pole? Give your answer to the nearest metre.

2. Jasmine is driving from east to west on a straight section of road. When she is at Madstowe, the town of Killingsworth is 40 km away and on a bearing of 345°. When she arrives in Lamport, Killingsworth is on a bearing of 080°. Find the distance from Lamport to Killingsworth, to the nearest kilometre.

3. Two girls are on opposite sides of a tower which is 17 metres tall. The girls measure the angle of elevation of the top of the tower as 32°15′ and 41°12′. How far apart are the two girls, to the nearest metre?

4. Jack is at A and the bearing of D from A is 130°. He then drives 23 km south to B which is directly west of D and then drives to C which is south of B. The bearing of D from C is 042°. What is the distance from D to C, to the nearest kilometre?

5. From the top of a shorter building the angle of elevation of the top of a taller building is 42° and the angle of depression of its base is 57°. If the buildings are 24 metres apart, find the heights of both buildings, to the nearest metre.

6. CHALLENGE Rose and Andrew are 100 metres apart and facing each other. They look up to see a kite. From Rose, the angle of elevation of the kite is 65°15′ and from Andrew it is 76°54′. How high is the kite, to the nearest metre?

Answers pages 157–158

KEY SKILL

47 Trigonometry: Two triangles B

HINTS

- Make sure you know the **Trigonometry Hints** from Key Skills 40 to 45 on pages 100, 102, 104, 106, 108 and 110.

Examples

Linda is standing on the top of an 80 metre-high cliff and observes a swimmer at an angle of depression of 48°. A minute later the swimmer has moved towards the cliff and the angle of depression is now 61°. How far has the swimmer travelled, to the nearest metre?

Solution

Let the 1st distance be x.

$$\frac{80}{x} = \tan 48^\circ$$
$$x = \frac{80}{\tan 48^\circ}$$

Let the 2nd distance be y.

$$\frac{80}{y} = \tan 61^\circ$$
$$y = \frac{80}{\tan 61^\circ}$$

$$\text{Difference} = \frac{80}{\tan 48^\circ} - \frac{80}{\tan 61^\circ}$$
$$= 28 \text{ (nearest whole)} \;\therefore\; 28 \text{ metres}$$

FOCUS on ...

1. The question: Asks you to find the distance the swimmer has travelled.
2. The information: Gives you the height of the cliff and the two angles of depression.
3. Your working: Draw a triangle to represent the information. Using alternate angles and parallel lines, the angles in the triangles can be recorded. Consider the two triangles to find two different distances. Let the first distance = x, and the later distance = y. In both cases, from the given angle, the sides are the opposite and adjacent. This means you will use tan:

$$\frac{80}{x} = \tan 48^\circ \qquad \frac{80}{y} = \tan 61^\circ$$
$$x = \frac{80}{\tan 48^\circ} \qquad y = \frac{80}{\tan 61^\circ}$$

You subtract to find the distance travelled:

$$\text{Distance} = \frac{80}{\tan 48^\circ} - \frac{80}{\tan 61^\circ}$$
$$= 27.687\,599\,43\ldots$$
$$= 28 \text{ (nearest whole)}$$

4. Your answer: Make sure you write the correct units: The swimmer travelled 28 m.

A straight road runs east–west. Tom is driving along the road and at A notices the bearing of a tower is 330°. After more driving, Tom reaches B and the tower is on a bearing of 345°. After another 10 km Tom is directly south of the tower. Find the distance from A to B, to the nearest kilometre.

Solution:

Let the distances be x, y.

$$\frac{y}{10} = \tan 75^\circ$$
$$y = 10 \times \tan 75^\circ$$

Now, $\dfrac{y}{x + 10} = \tan 60^\circ$

$$x + 10 = \frac{y}{\tan 60^\circ}$$

$$x = \frac{10 \times \tan 75^\circ}{\tan 60^\circ} - 10 = 11.547\,005\,38\ldots$$
$$= 12 \text{ (nearest whole)} \quad \therefore\; 12 \text{ km.}$$

FOCUS on ...

1. The question: Asks you to find the distance AB.
2. The information: Gives you the bearings of a tower from A and B on an east-west road, and the distance B is from a position (C) directly south of the tower.
3. Your working: Draw a diagram to represent the information and locate C which is the closest point on the road from the tower. Calculate the angles inside the triangle. Let AB = x and TC = y. By considering △BCT, find the distance TC:

$$\frac{y}{10} = \tan 75^\circ$$
$$y = 10 \times \tan 75^\circ$$

Now use △ACT, to find AC, and therefore, AB:

$$x + 10 = \frac{y}{\tan 60^\circ}$$
$$x = \frac{10 \times \tan 75^\circ}{\tan 60^\circ} - 10$$
$$= 11.547\,005\,38\ldots$$
$$= 12 \text{ (nearest whole)}$$

4. Your answer: Make sure you write the correct units: The distance is 12 km.

Now try these!

1. Pim stood at a window 45 metres from street level and observed her friend Tony walking away from her at an angle of depression of 60°. After half a minute, the angle of depression of Tony was 40°. How far had he walked in that time, to the nearest metre?

2. A straight road runs east–west. Eric is driving along the road and when he reaches a rest area he calculates that the bearing of Gandra is 280°. Later he stops at a service station and determines that the bearing of Gandra is 340°. After another 20 km he reaches the point where he is directly south of Gandra. How far is it from the rest area to the service station, to the nearest kilometre?

3. A flagpole is placed on the top of a building. At a point on street level 48 metres from the base of the building, the angle of elevation of the top of the pole is 52° and the top of the building is 47°. Find the height of the flagpole, to the nearest metre.

4. Ben and Ken are attending a musical. Ben's seat is at floor level and he looks down at an angle of 16° to the orchestra pit to see the conductor. Ken is sitting directly above Ben on the next level and he looks down 46° to see the conductor. If the horizontal distance from Ben to the conductor is 8 metres, how far above Ben's seat is Ken's seat, to the nearest metre?

5. From where she is standing, Bella calculates the angle of elevation from the ground to the top of a tree to be 47°. When she walks 20 metres closer to the tree the angle is 63°. What is the height of the tree, to the nearest metre?

6. CHALLENGE Wade is driving south at a constant speed of 80 km/h. At 9 am the bearing of a tower is 170° and at 9:45 am the bearing is 140°. At what time will Wade be west of the tower?

Answers pages 158–159

REVISION TEST 13 Level of difficulty—Average

1

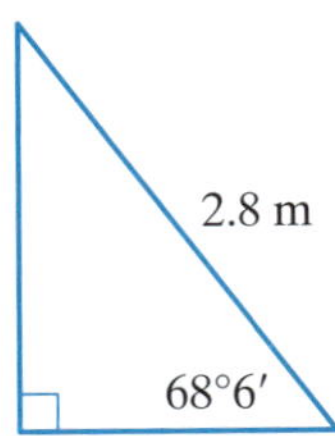

A ladder is 2.8 metres long and leans against a wall, making an angle of 63°6′ with the ground. How far is the base of the ladder from the wall? Give your answer correct to 2 decimal places.

2 Find the size of each angle of a triangle with the dimensions 8 cm, 8 cm and 10 cm. Give your answer to the nearest degree.

3

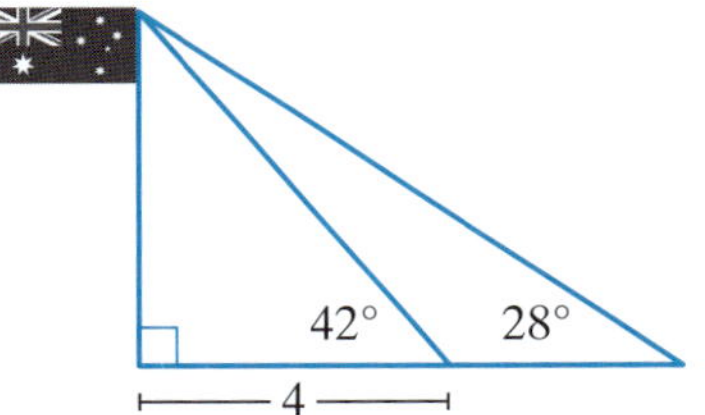

The shadow of a flagpole when the angle of elevation of the sun is 42° is 4 metres long. How much longer is the shadow when the sun is at 28°? Give your answer correct to 3 significant figures.

4 Two vertical poles are 40 metres apart. The angle of depression of the top of the smaller pole from the top of the taller pole is 26°15′. If the shorter pole is 7 metres high, what is the height of the taller pole, to the nearest metre?

5 A ladder with its foot in a narrow alley makes an angle of 50° with the ground when it rests on a building on one side of the street. When the ladder is leaned to the building on the other side of the alley it forms an angle of 60° with the ground. If the ladder is 3.1 metres long, how wide is the alley, correct to 2 decimal places?

6 A rectangular sign is placed on the top of a building. At a point on street level 60 metres from the base of the building, the angle of elevation of the top of the sign is 48° and the top of the building is 43°. Find the height of the sign, to the nearest metre.

Answers page 159

REVISION TEST 14 Level of difficulty—Challenging

1. Paxton is 26 km south-east of Quinn and Robertson is 32 km north-east of Quinn. What is the bearing of Robertson from Paxton?

2. 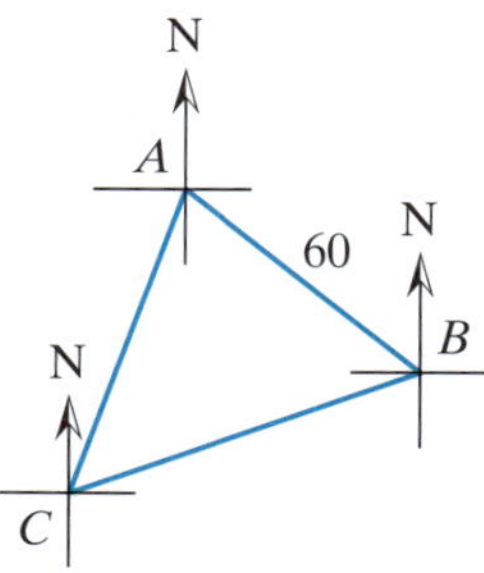

The distance from A to B is 60 km. The bearing of B from A is 130°, the bearing of C from B is 240° and the bearing of A from C is 040°. Find the distance from B to C, to the nearest kilometre.

3. Alec and Belinda are standing 80 metres apart and are facing each other. When they look up they see a seagull at an angle of elevation from Alec of 20° and at 43° from Belinda. What is the height of the seagull, to the nearest metre?

4. A plane is flying at a constant height of 6000 metres and a constant speed of 480 km/h. The angle of depression of a lake is 42°. How long, to the nearest second, will it take for the plane to be directly over the lake?

5. Two boats, A and B, are anchored 70 km apart. The bearing of B from A is 050°. At 2 pm B sails on a bearing of 200° at a constant speed of 20 km/h. The boat reaches point C, the position on its course where it is at its closest to A. Find the distance AC and the bearing of C from A.

6. Myles is driving west at a constant speed of 80 km/h. At 2:30 pm the bearing of a mountain peak is 330° and at 4:45 pm the bearing is 350°. At what time will Myles be south of the mountain peak?

Answers page 160

KEY SKILL

STATISTICS AND PROBABILITY

48 Statistics: Box plots

HINTS

- A box plot is a statistical diagram showing five measures of statistics: lower extreme (lowest score), lower quartile, median, upper quartile and upper extreme (highest score), e.g. Draw a box plot for the scores 7, 9, 11, 12, 14, 14, 15.

7 8 9 10 11 12 13 14 15

Examples

A science teacher recorded the results of six students on a quiz: 3, 6, 8, 10, 10, 12. Two students were absent, but they later completed the quiz and their marks were recorded. The teacher completed a box plot of the eight results:

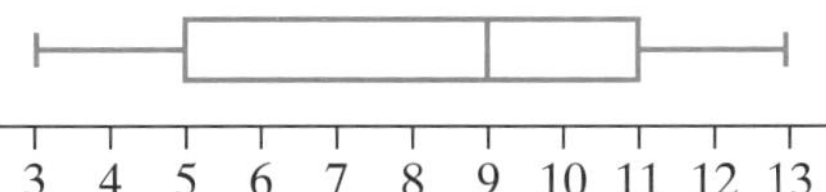

What are the two new results?

Solution

The upper extreme was 12 and is now 13, so there is a new score of 13.

This will change the upper quartile.

The lower quartile was 6 and is now 5, so there is a new score of 4.

The 8 scores are 3, 4, 6, 8, 10, 10, 12, 13.

$\therefore$ the new scores are 4 and 13.

FOCUS on ...

1. The question: Asks you to find the two new results.
2. The information: Gives you the six results and the box plot of eight results.
3. Your working: You need to compare the box plots of the six scores and the box plot of the eight scores:

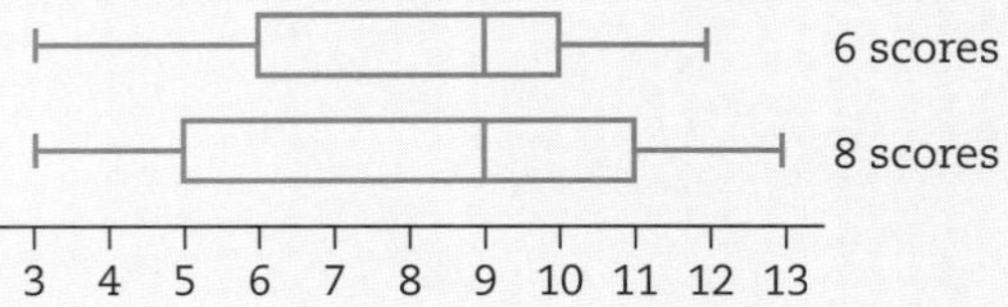

The highest score (upper extreme) has changed: this means one of the new scores is 13.

The lower quartile has changed from 6 to 5: the scores were 3, 6, 8, ... and now they must be 3, x, 6, 8, ... As the new lower quartile is 5, then the middle of x and 6 has to be 5. This means $x = 4$.

The eight scores are 3, 4, 6, 8, 10, 10, 12, 13.

4. Your answer: The new scores are 4 and 13.

A set of scores in ascending order is listed: a, 3, b, 5, c, 7, d, 12, 14. Brittany draws a box plot for the data:

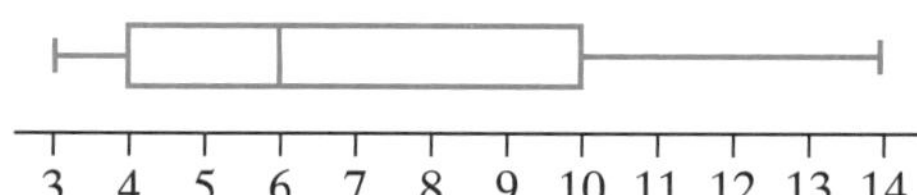

What are the values of a, b, c and d?

Solution

Lower extreme is 3. $\therefore a = 3$

Median is 6. $\therefore c = 6$

Lower quartile is 4. $\therefore b = 5$

Upper quartile is 10. $\therefore d = 8$

The scores are 3, 3, 5, 5, 6, 7, 8, 12, 14

$\therefore a = 3, b = 5, c = 6, d = 8$.

FOCUS on ...

1. The question: Asks you to find the values of the pronumerals.
2. The information: Gives you a set of scores, in ascending order, including four unknown scores as well as a box plot.
3. Your working: You need to compare the list of scores to the box plot.

Lower extreme = 3 means there is another 3 in the list: this means $a = 3$.

As the median = 6, and there are 9 scores, the middle (5th) number must be 6: this means $c = 6$.

You can now rewrite the scores:

3, 3, b, 5, 6, 7, d, 12, 14.

Lower quartile = 4, so middle of 3 and b is 4: this means that $b = 5$.

Upper quartile = 10, so middle of d and 12 is 10: this means that $d = 8$.

The scores are 3, 3, 5, 5, 6, 7, 8, 12, 14.

4. Your answer: $a = 3, b = 5, c = 6, d = 8$.

Now try these!

1. A history teacher recorded the results for five students on a quiz: 6, 8, 10, 7, 5. Two more students were absent, but they completed the quiz later. The teacher completed a box plot for the seven students:

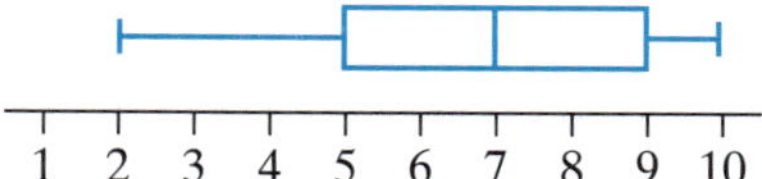

What were the two new results?

2. A set of scores in ascending order is listed: a, 8, b, 10, c, 14, d, 16. Jason draws a box plot for the data:

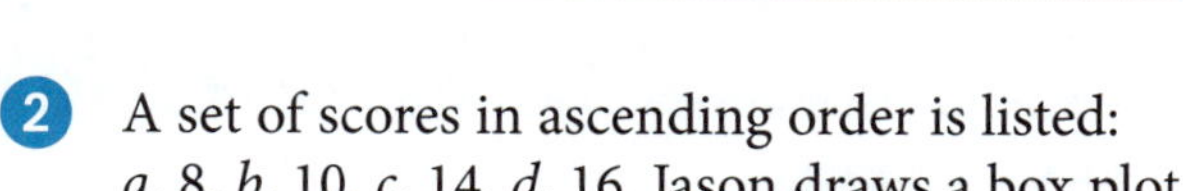

What are the values of a, b, c and d?

3. The following marks were scored in a test: 8, 12, 10, 16, 14, 18, 10, 15, 9, 12
Two of the scores were ignored when the box plot was drawn.

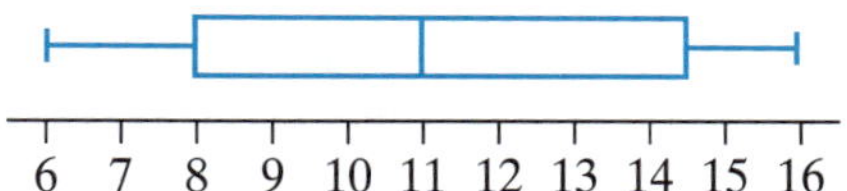

What two scores were ignored?

4. There are 28 students in Mr Ireland's maths class and each student completed a test marked out of 40. The median mark was 60% and 7 students scored 50% or less. The interquartile range was 12, the top student scored 95% and the range of marks for the class was 32. Complete the box plot.

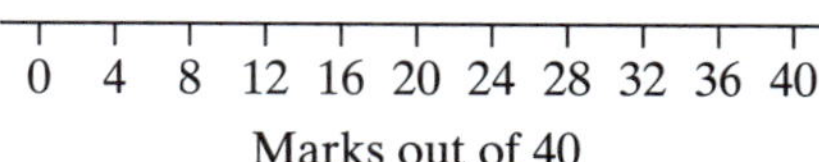

5. Twenty students completed two science tests and their results are shown in the box plots.

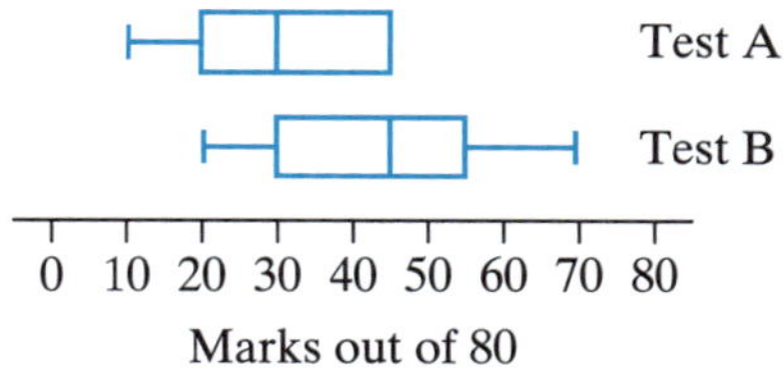

What is the smallest number of students who improved their results from Test A to Test B?

6. CHALLENGE Use the box plot to find the values of a, b, c, d and e in the table, given $a < b < c < d < e$.

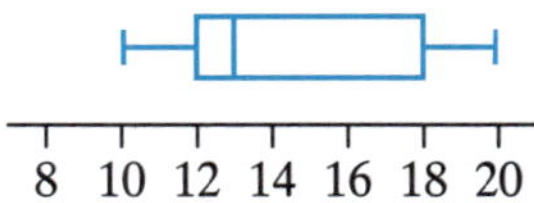

Scores	a	b	c	d	e
Frequency	2	8	4	4	2

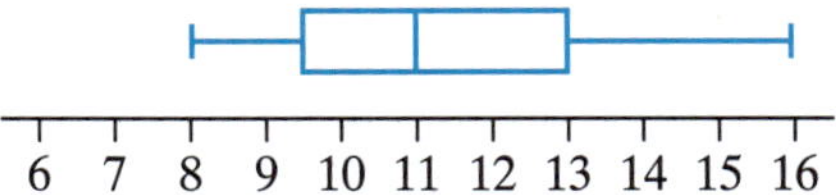

Answers pages 160–161

KEY SKILL

49 Probability: Tree diagrams

HINTS

- A probability tree, or tree diagram, is used to display the outcomes of a multistage experiment.
- The probability of each step can be written on each branch.
- Multiply when the outcomes occur at the same time/consecutively (uses the word 'and'), and add if the outcomes are alternatives (uses the word 'or'),

  e.g. Use the probability tree to find the probability of BB.

  $\therefore$ P(BB) $= \frac{3}{5} \times \frac{2}{4} = \frac{3}{10}$

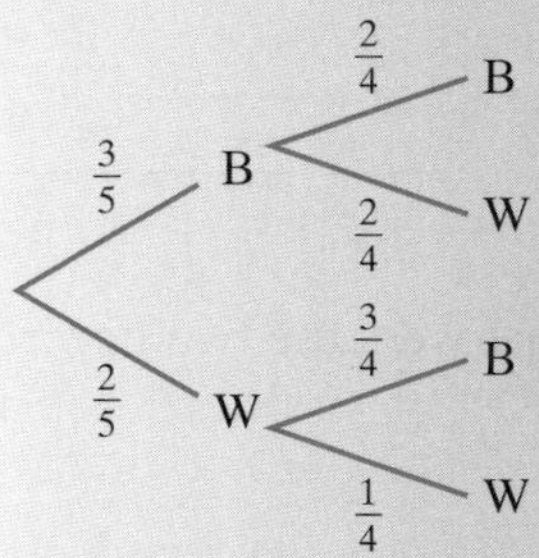

Examples

Of the 8 marbles in a bag, exactly 3 are green. If Graeme reaches into the bag and pulls out 2 marbles, what is the probability that they are both green?

Solution

P(GG)

$= \frac{3}{8} \times \frac{2}{7}$

$= \frac{3}{28}$

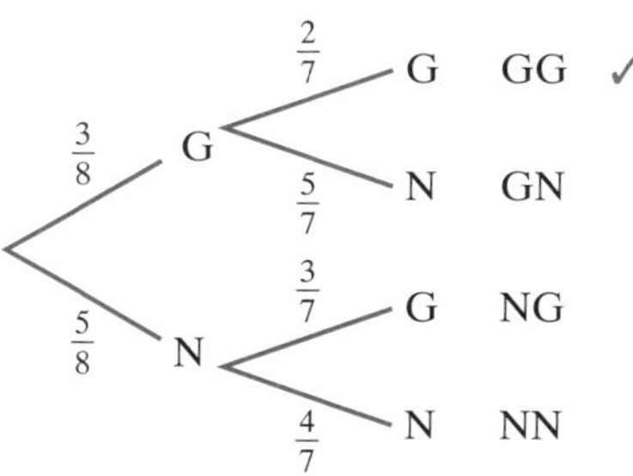

$\therefore$ the probability of both green is $\frac{3}{28}$.

FOCUS on ...

1. The question: Asks you to find the probability that Graeme chooses 2 green marbles.
2. The information: Gives you the number of marbles in a bag, the number that are green and that 2 marbles are to be chosen, without replacement.
3. Your working: You need to draw a tree diagram to show the probabilities of each alternative.

   P(first marble green) $= \frac{3}{8}$ and, if a green was chosen, then P(second marble green) $= \frac{2}{7}$.

   You multiply the probabilities:

   P(GG) $= \frac{3}{8} \times \frac{2}{7} = \frac{3}{28}$
4. Your answer: The probability is $\frac{3}{28}$.

A box contains 2 red balls and 3 blue balls. A ball is chosen at random and removed. It is then replaced with 2 balls of the opposite colour. What is the probability of choosing balls of different colours?

Solution

P(RB) + P(BR)

$= \frac{2}{5} \times \frac{5}{6} + \frac{3}{5} \times \frac{4}{6}$

$= \frac{22}{30}$

$= \frac{11}{15}$

2/5 R — 1/6 R: RR; 5/6 B: RB ✓

3/5 B — 4/6 R: BR ✓; 2/6 B: BB

$\therefore$ probability of different colours is $\frac{11}{15}$.

FOCUS on ...

1. The question: Asks you to find the probability that two chosen balls are different colours.
2. The information: Gives you the number of balls in a box and that when one is chosen it is removed and replaced with two balls of the opposite colour.
3. Your working: You need to draw a tree diagram to show the probabilities of each alternative.

   P(first ball red) $= \frac{2}{5}$; red ball removed and replaced with 2 blue balls, then P(second ball blue) $= \frac{5}{6}$.

   Similarly, P(first ball blue) $= \frac{3}{5}$; blue ball removed and replaced with 2 red balls, then P(second ball red) $= \frac{4}{6}$.

   You multiply the probabilities and then add:

   P(RB or BR) $= \frac{2}{5} \times \frac{5}{6} + \frac{3}{5} \times \frac{4}{6} = \frac{22}{30} = \frac{11}{15}$.
4. Your answer: The probability is $\frac{11}{15}$.

Now try these!

1. There are 10 balls in a bag. Three of the balls are blue and the remainder are red. Two balls are selected from the bag at random. Draw a tree diagram to list all possible outcomes. What is the probability that the balls are different colours?

2. A bag contains 4 black discs and 2 white discs. A disc is chosen at random and the colour noted. It is placed back into the bag and another disc of the opposite colour is also added to the bag. When a second disc is chosen what is the probability that it is the same as the first disc?

3. Angie buys 3 tickets in a raffle of 20 tickets. There are two prizes. Draw a tree diagram to list all possible outcomes. What is the probability that Angie wins second prize?

4. A chessboard has 32 white squares and 32 black squares. Meg chooses two squares at random. What is the probability that at least one of the squares is white?

5. Jack and Ryan are playing in a chess competition. They play two games and each has an equal chance of winning the first game. If Jack wins the first game, the probability of him winning the second game increases to 0.7. If Jack loses the first game, the probability of him winning the second game drops to 0.4. What is the probability that Jack wins exactly one game?

6. **CHALLENGE** Isabella and Madison are playing a game, taking turns rolling two dice. The game is won by the first player to roll a double six. Isabella has the first roll. What is the probability that she wins the game on her second turn, to 2 decimal places?

Answers page 161

KEY SKILL

50 Probability: Conditional probability

HINTS

- The conditional probability of an event A is the probability that it will happen given that B has already occurred and it is written as P(A|B).
- $P(A|B) = \frac{P(A \cap B)}{P(B)}$ = P(both A and B) ÷ P(B)

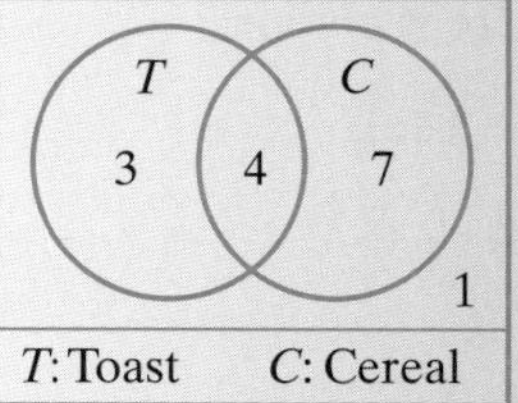

e.g. The Venn diagram shows the number of people who have toast and/or cereal for breakfast. What is the probability that a person has toast, given that they had cereal?

P(both) $= \frac{4}{15}$, P(C) $= \frac{11}{15}$, P(toast | cereal) $= \frac{4}{15} \div \frac{11}{15} = \frac{4}{11}$

Examples

Rachael rolled two dice and added the numbers. What is the probability that the total is greater than 9, given that at least one of the dies is showing a 6?

Solution

From {(1, 6), (2, 6), (3, 6), (4, 6), (5, 6), (6, 6), (6,1), (6, 2), (6, 3), (6, 4), (6, 5)}, need to choose {(4, 6), (5, 6), (6, 6), (6, 4), (6, 5)}.

P(sum > 9|at least one is 6)

= P(shows a 6 & total > 9) ÷ P(shows at least a 6)

$= \frac{5}{36} \div \frac{11}{36}$

$= \frac{5}{11}$ $\quad \therefore$ probability is $\frac{5}{11}$.

FOCUS on…

1. The question: Asks you to find the probability that the total is more than 9, given at least one die shows a 6.
2. The information: Gives you that two dice are rolled and the total is calculated, and one shows a 6.
3. Your working: You need to list the number of ways 6 can be on at least one die, and then count how many of these add to a score higher than 9.

   P(sum is greater than 9, given that at least one is a 6) = P(sum > 9|at least one is a 6)

   $= \frac{5}{11}$
4. Your answer: The probability is $\frac{5}{11}$.

A two-way table is used to summarise 16 teams rehiring the coach and/or making the playoffs for another season.

		Team making playoffs	
		Yes	No
Coach	Re-hired	6	3
	Sacked	2	5

What is the probability that a coach is rehired, given that the team does not make the playoffs?

Solution

P(re-hired|no playoffs)

= P(re-hired & no playoffs) ÷ P(no playoffs)

$= \frac{3}{16} \div \frac{8}{16} = \frac{3}{8}$ $\quad \therefore$ probability is $\frac{3}{8}$.

FOCUS on…

1. The question: Asks you to find the probability that a coach is rehired, given that the team does not make the playoffs.
2. The information: Gives you a two-way table which can be used to find different probabilities.
3. Your working: You need to find the number of teams who did not make play-offs but rehired their coach: 3 out of 16. Also the number of teams who did not make playoffs: 8 out of 16.

   P(rehired|no playoffs)

   = P(rehired and no playoffs) ÷ P(no playoffs)

   $= \frac{3}{16} \div \frac{8}{16}$

   $= \frac{3}{8}$
4. Your answer: The probability is $\frac{3}{8}$.

Now try these!

1. A family has two children. It is known that one of the children is a girl. What is the probability that the other child is a boy?

A survey found the way students had travelled to school this morning and the results are shown in the table.

		Mode of transport			
		Bus	**Car**	**Walk**	**Bike**
School years	**7, 8**	24	44	20	12
	9, 10	32	38	22	8
	11, 12	28	50	16	6

Use the table to answer questions 2 and 3.

2. What is the probability that a student travelled by car, given that he or she is in Years 9 or 10?

3. What is the probability that a student is in Years 7 or 8, given that he or she walked to school?

Otis surveyed a group of senior students and the diagram shows the number of students studying Biology, Chemistry and Physics.

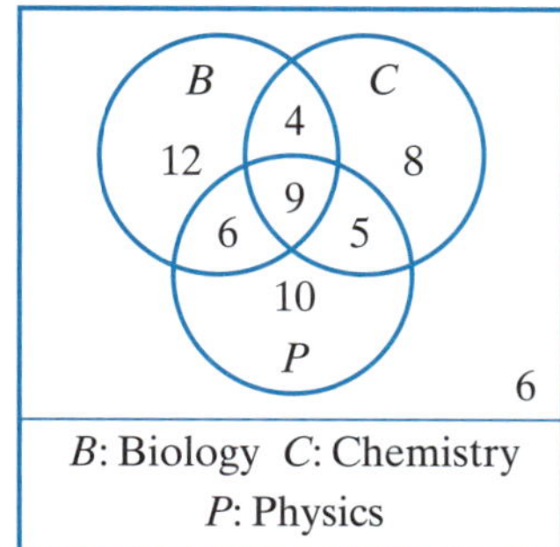

B: Biology *C*: Chemistry *P*: Physics

Use the diagram to answer questions 4 and 5.

4. What is the probability that a student studies Chemistry, given that he or she studies Biology?

5. What is the chance that a student who is not studying Chemistry does not study Physics?

6. CHALLENGE At a secondhand book sale there are 2000 books. There are 800 children's books and the remainder are adults' books. Half of the 1100 non-fiction books are for adults. A book is chosen at random. What is the probability that it is fiction, given that it is a children's book?

Answers page 162

REVISION TEST 15 Level of difficulty—Average

1 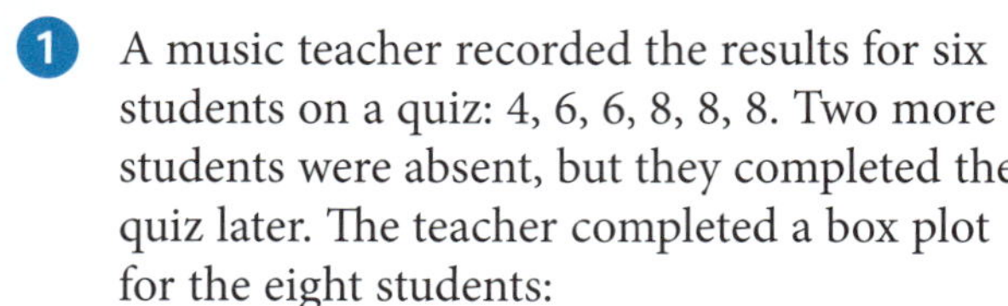

A music teacher recorded the results for six students on a quiz: 4, 6, 6, 8, 8, 8. Two more students were absent, but they completed the quiz later. The teacher completed a box plot for the eight students:

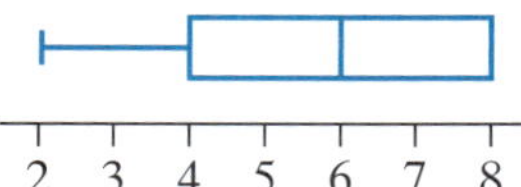

What were the two new results?

2 There are eight balls in a box. Three of the balls are green and the remainder are red. Two balls are selected from the bag at random. Draw a tree diagram to list all possible outcomes. What is the probability that the balls are different colours?

3 A bag contains 26 cards marked with the letters A to Z. A card is chosen at random from the bag and discarded. A second selection is then made. What is the probability that a vowel was chosen on both occasions?

4 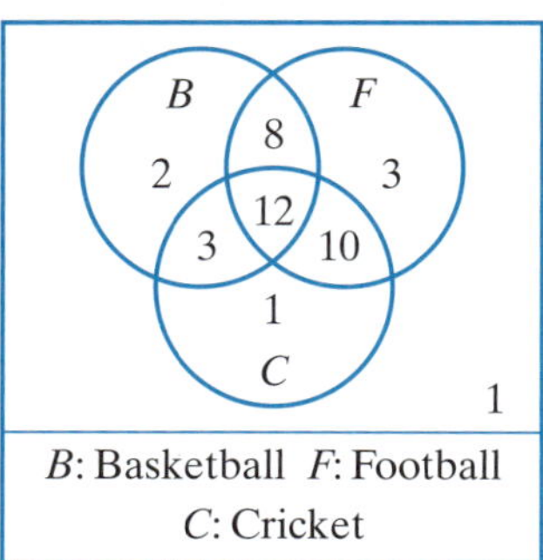

The Venn diagram shows the results of a survey to find the sporting interests of a group of people. If a person was chosen at random, what is the probability that he or she enjoys cricket but not football?

5 In a sample of 40 African national flags, Ethan found that 35 use the colour green, 32 use the colour red and 28 use both colours. Use a two-way table to find the probability that a randomly chosen flag does not use green or red.

6 The data from a survey of 100 people in a country town showed that 34 were bowling-club members, 28 were golf-club members and 52 were not members of either club. Use a Venn diagram to find the probability that a randomly chosen person was a member of the bowling club given that he or she was not a member of the golf club.

Answers pages 162–163

REVISION TEST 16 Level of difficulty—Challenging

1 Use the box plot to find the values of p, q, r, s and t in the table, given $p < q < r < s < t$.

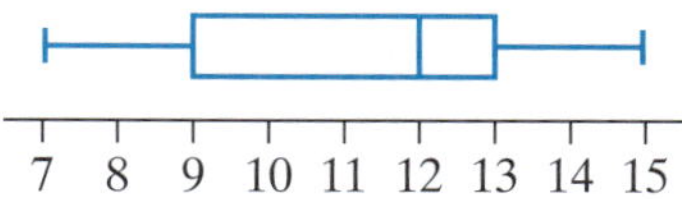

Scores	p	q	r	s	t
Frequency	1	2	1	3	1

2 A group of 50 students, which included 20 boys, were surveyed. Fifteen students wore glasses, and 8 boys in the group wore glasses. If a student was chosen at random, what was the probability that the student was a girl who did not wear glasses?

3 Two identical biased coins are tossed together, and the outcome is recorded. After a large number of trials it is observed that the probability that both coins land showing tails is 0.16. What is the probability that both coins land showing heads?

4 Jack and Jill play two games of tiddlywinks. Jill is twice as likely to win a game as Jack. Complete the probability tree and find the probability that they will win one game each.

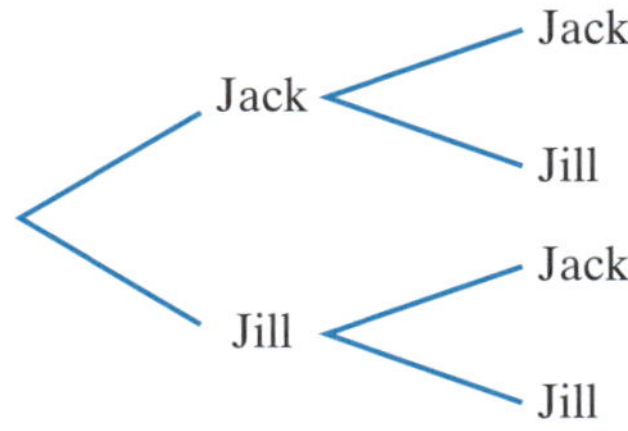

5 Eight out of the 18 balls in a box are green. Another four balls are red and the remainder are blue. Two balls are chosen at random. What is the probability that neither ball is blue?

6 In a group of 100 Year 12 students, 38 study Music, 32 study Physics, 35 study Japanese, 11 study Music and Physics, 16 study Physics and Japanese, 8 study Music and Japanese, and 6 study all three subjects. Use a Venn diagram to find the probability that a randomly chosen student studies Japanese, given that he or she does not study Physics.

Answers page 163

Worked solutions

Key Skill  (pages 10–11)

1. **\$1331.20**
   Total pay $= 25.6 \times 28 + 25.6 \times 1.5 \times 8 + 25.6 \times 2 \times 6 = 1331.2$
   ∴ her total pay was \$1331.20.
2. **\$21**
   Let her hourly rate be x.
   $756 = 18 \times x + 4 \times 1.5 \times x + 6 \times 2 \times x$
   $= 18x + 6x + 12x$
   $= 36x$
   $x = \frac{756}{36}$
   $= 21$
   ∴ her hourly rate was \$21.
3. **\$4575**
   Total pay
   $= 75 \times 42 + 75 \times 1.5 \times 6 + 75 \times 2 \times 5$
   $= 4575$
   ∴ his total pay was \$4575.
4. **4**
   Let her double-time hours be x.
   $813.2 = 21.4 \times 18 + 21.4 \times 1.5 \times 8 + 21.4 \times 2 \times x$
   $= 642 + 42.8x$
   $42.8x = 813.2 - 642$
   $42.8x = 171.2$
   $x = \frac{171.2}{42.8}$
   $= 4$
   ∴ Tara worked 4 hours at double time.
5. **Jordan by \$39.60**
   Damian's pay
   $= 13.2 \times 10 + 13.2 \times 1.5 \times 6$
   $= 250.8$
   Jordan's pay
   $= 13.2 \times 8 + 13.2 \times 2 \times 7$
   $= 290.4$
   Difference $= 290.4 - 250.8$
   $= 39.6$
   ∴ Jordan is paid \$39.60 more than his brother.
6. **\$289.50**
   Let his hourly rate be x.
   $492.15 = 12 \times x + 9 \times 1.5 \times x$
   $= 12x + 13.5x$
   $= 25.5x$
   $x = \frac{492.15}{25.5}$
   $= 19.3$
   ∴ his hourly rate is \$19.30.
   Next pay $= 19.3 \times 9 + 19.3 \times 2 \times 3$
   $= 289.5$
   ∴ Leonard will earn \$289.50.

Key Skill  (pages 12–13)

1. **\$6019.25**
   Taxable income $= 46\,920 - 2390$
   $= 44\,530$
   Tax payable $= 3572 + 0.325 \times 7530$
   $= 6019.25$
   ∴ Amelia will pay \$6019.25.
2. **\$1549.80**
   Total fortnightly tax paid
   $= 788 \times 26$
   $= 20\,488$
   Taxable income $= 86\,300 - 2540$
   $= 83\,760$
   Tax payable $= 17\,547 + 0.37 \times 3760$
   $= 18\,938.2$
   Refund $= 20\,488 - 18\,938.2$
   $= 1549.8$
   ∴ Cooper will receive a refund of \$1549.80.
3. **\$14 695.45**
   Total deductions $= 670 + 504$
   $= 1174$
   Taxable income $= 72\,400 - 1174$
   $= 71\,226$
   Tax payable
   $= 3572 + 0.325 \times 34\,226$
   $= 14\,695.45$
   ∴ Mia will pay \$14 695.45.
4. **\$11.50 more to ATO**
   Total income $= 38\,600 + 16\,460$
   $= 55\,060$
   Tax payable
   $= 3572 + 0.325 \times 18\,060$
   $= 9441.5$
   Difference $= 9441.5 - 9430$
   $= 11.5$
   ∴ Heidi still needs to pay \$11.50 more.
5. **23.5% and 13.6%**
   Jayden's tax payable
   $= 17\,547 + 0.37 \times 9400$
   $= 21\,025$
   Jayden's tax % $= \frac{21\,025}{89\,400} \times 100\%$
   $= 23.5$ (1 dec. pl.)
   Bella's taxable income $= 89\,400 \div 2$
   $= 44\,700$
   Bella's tax payable
   $= 3572 + 0.325 \times 7700$
   $= 6074.5$
   Bella's tax % $= \frac{6074.5}{44\,700} \times 100\%$
   $= 13.6$ (1 dec. pl.)
   ∴ Jayden pays 23.5% and Bella pays 13.6% of their income in tax.
6. **\$50.40**
   Annual income $= 3200 \times 26$
   $= 83\,200$
   Tax payable $= 17\,547 + 0.37 \times 3200$
   $= 18\,731$
   Increased income $= 83\,200 \times 1.025$
   $= 85\,280$
   New tax payable
   $= 17\,547 + 0.37 \times 5280$
   $= 19\,500.6$
   Extra tax $= 19\,500.60 - 18\,731$
   $= 769.6$
   Extra fortnightly tax $= 29.6$
   Extra weekly income
   $= 3200 \times 0.025$
   $= 80$
   Extra in pay packet $= 80 - 29.6$
   $= 50.4$
   ∴ Ashley will receive an extra \$50.40.

Key Skill  (pages 14–15)

1. **\$94 600**
   Xavier's income was between \$80 000 and \$180 000.
   Let his income be x.
   $22\,949 = 17\,547 + 0.37 \times (x - 80\,000)$
   $5402 = 0.37(x - 80\,000)$
   $x - 80\,000 = \frac{5402}{0.37}$
   $= 14\,600$
   $x = 94\,600$
   ∴ Xavier had a taxable income of \$94 600.
2. **\$78 450**
   Total fortnightly tax paid
   $= 670 \times 26$
   $= 17\,420$
   Tax payable from table
   $= 17\,420 - 896.75$
   $= 16\,523.25$
   His income was between \$37 000 and \$80 000.
   Let his income be x.
   $16\,523.25 = 3572 + 0.325 \times (x - 37\,000)$
   $12\,951.25 = 0.325(x - 37\,000)$
   $x - 37\,000 = \frac{12\,951.25}{0.325}$
   $= 39\,850$

$x = 39\,850 + 37\,000$
$= 76\,850$
Gross earnings $= 76\,850 + 1600$
$= 78\,450$
$\therefore$ Charlie's gross earnings were $78 450.

3. **$130 520**
Jordan's income was between $80 000 and $180 000.
Let his income be x.
$34\,567 = 17\,547 + 0.37 \times (x - 80\,000)$
$17\,020 = 0.37(x - 80\,000)$
$x - 80\,000 = \dfrac{17\,020}{0.37}$
$= 46\,000$
$x = 46\,000 + 80\,000$
$= 126\,000$
Gross income $= 126\,000 + 4520$
$= 130\,520$
$\therefore$ Jordan had a gross income of $130 520.

4. **$25 800**
Total fortnightly tax paid $= 60 \times 26$
$= 1560$
Tax payable from table $= 1560 - 116$
$= 1444$
His income was between $18 200 and $37 000.
Let his income be x.
$1444 = 0.19 \times (x - 18\,200)$
$x - 18\,200 = \dfrac{1444}{0.19}$
$= 7600$
$x = 7600 + 18\,200$
$= 25\,800$
$\therefore$ Braxton's taxable income was $25 800.

5. **$89 000**
Total fortnightly tax paid
$= 740 \times 26$
$= 19\,240$
Tax payable from table
$= 19\,240 + 1560$
$= 20\,800$
Her income was between $80 000 and $180 000.
Let her income be x.
$20\,800 = 17\,547 + 0.37 \times (x - 80\,000)$
$3253 = 0.37 \times (x - 80\,000)$
$x - 80\,000 = \dfrac{3253}{0.37}$
$= 8791.89$ (2 dec pl)
$x = 8791.89 + 80\,000$
$= 88\,791.89$
$= 89\,000$ (nearest thousand)
$\therefore$ Brodie's taxable income was $89 000.

6. **$54 313.85**
Cyril's tax payable
$= 17\,547 + 0.37 \times (82\,300 - 80\,000)$
$= 18\,398$
As $18\,398 \div 2 = 9199$, Cedric paid $9199 tax.
Cedric's income was between $37 000 and $80 000.
Let his income be x.
$9199 = 3572 + 0.325 \times (x - 37\,000)$
$5627 = 0.325(x - 37\,000)$
$x - 37\,000 = \dfrac{5627}{0.325}$
$= 1731.846\,15\ldots$
$= 17\,313.85$ (2 dec. pl.)
$x = 54\,313.85$
$\therefore$ Cedric had a taxable income of $54 313.85.

Key Skill  4 (pages 16–17)

1. **$570**
Subs. $P = 3800, r = 0.0375, n = 4$
$I = Prn$
$= 3800 \times 0.0375 \times 4$
$= 570$
$\therefore$ Chelsea earned simple interest of $570.

2. **$4620**
Subs. $P = 35\,000, r = 0.0055, n = 24$
$I = Prn$
$= 35\,000 \times 0.0055 \times 24$
$= 4620$
$\therefore$ simple interest of $4620 will be earned.

3. **$816**
Subs. $P = 8500, r = 0.016, n = 6$
$I = Prn$
$= 8500 \times 0.016 \times 6$
$= 816$
$\therefore$ Caitlin will earn simple interest of $816.

4. **$2430**
Subs. $P = 27\,000, r = 0.005, n = 18$
$I = Prn$
$= 27\,000 \times 0.005 \times 18$
$= 2430$
$\therefore$ Ruby earns simple interest of $2430.

5. **Ewan by $100**
Ewan: Subs. $P = 10\,000, r = 0.065, n = 5$
$I = Prn$
$= 10\,000 \times 0.065 \times 5$
$= 3250$
Oliver: Subs. $P = 10\,000, r = 0.0525, n = 6$:
$I = Prn$
$= 10\,000 \times 0.0525 \times 6$
$= 3150$
Difference $= 3250 - 3150$
$= 100$
$\therefore$ Ewan earns $100 more than Oliver.

6. **$829.12**
Subs. $P = 4500, r = 0.0315, n = 3$
$I = Prn$
$= 4500 \times 0.0315 \times 3$
$= 425.25$
Total $= 4500 + 425.25$
$= 4925.25$
Subs. $P = 4925.25, r = 0.041, n = 2$:
$I = Prn$
$= 4925.25 \times 0.041 \times 2$
$= 403.8705$
$= 403.87$ (2 dec pl)
Total $= 425.25 + 403.87$
$= 829.12$
$\therefore$ total interest of $829.12

Key Skill 5 (pages 18–19)

1. **$3600**
Subs. $I = 1008, r = 0.04, n = 7$
$I = Prn$
$1008 = P \times 0.04 \times 7$
$1008 = 0.28P$
$P = \dfrac{1008}{0.28}$
$= 3600$
$\therefore$ Morton invested $3600.

2. **5.4%**
Subs. $I = 1728, P = 6400, n = 5$
$I = Prn$
$1728 = 6400 \times r \times 5$
$1728 = 32\,000r$
$r = \dfrac{1728}{32\,000}$
$= 0.054$
Rate as % $= 0.054 \times 100\%$
$= 5.4\%$
$\therefore$ the interest rate was 5.4% p.a.

3. **20 years**
Subs. $I = 8000, P = 8000, r = 0.05$
$I = Prn$
$8000 = 8000 \times 0.05 \times n$
$8000 = 400n$
$n = \dfrac{8000}{400}$
$= 20$
$\therefore$ it would take Samuel 20 years.

4. $475

Grace: Subs. $I = 1900$, $P = 5000$, $n = 8$

$$I = Prn$$
$$1900 = 5000 \times r \times 8$$
$$1900 = 40\,000r$$
$$r = \frac{1900}{40\,000}$$
$$= 0.0475$$

Erik: Subs. $P = 5000$, $r = 0.0475$, $n = 10$

$$I = Prn$$
$$= 5000 \times 0.0475 \times 10$$
$$= 2375$$
$$\text{Difference} = 2375 - 1900$$
$$= 475$$

∴ Erik will earn $475 more interest.

5. 3 years

6% p.a. = 0.05% per month

Subs. $I = 1728$, $P = 9600$, $r = 0.005$

$$I = Prn$$
$$1728 = 9600 \times 0.005 \times n$$
$$1728 = 48n$$
$$n = \frac{1728}{48}$$
$$= 36$$

∴ Phoebe invested for 36 months, or 3 years.

6. 3% p.a. and 4% p.a.

Let rate on first account be x% p.a.

Let rate on second account be $(x + 1)$% p.a.

$$8000 \times \frac{x}{100} \times 2 = 320 + 2000 \times \frac{x+1}{100} \times 2$$
$$160x = 320 + 40x + 40$$
$$120x = 360$$
$$x = 3$$

∴ interest rates were 3% p.a. and 4% p.a.

Key Skill  (pages 20–21)

1. $96.69

$$\text{Deposit} = 0.2 \times 2590$$
$$= 518$$
$$\text{Balance owing} = 2072$$
$$\text{Interest} = 2072 \times 0.06 \times 2$$
$$= 248.64$$
$$\text{Total repayments} = 2320.64$$

Monthly repayments

$= 2320.64 \div 24$

$= 96.693\,3333\ldots$

$= 96.69$ (2 dec. pl.)

∴ the monthly instalment is $96.69.

2. 6% p.a.

$$\text{Deposit} = 6000$$
$$\text{Balance owing} = 48\,000$$
$$\text{Total instalments} = 1040 \times 60$$
$$= 62\,400$$
$$\text{Total interest} = 62\,400 - 48\,000$$
$$= 14\,400$$
$$I = Prn$$
$$14\,400 = 48\,000 \times r \times 5$$
$$14\,400 = 240\,000r$$
$$r = \frac{14\,400}{240\,000}$$
$$= 0.06$$

∴ Alyssa is charged interest of 6% p.a.

3. $1139.75

$$\text{Trade-in} = 5000$$
$$\text{Balance owing} = 47\,000$$
$$\text{Interest} = 47\,000 \times 0.041 \times 4$$
$$= 7708$$
$$\text{Total repayments} = 54\,708$$
$$\text{Monthly repayments} = 54\,708 \div 48$$
$$= 1139.75$$

∴ the monthly instalment is $1139.75.

4. 4% p.a.

$$\text{Deposit} = 3200$$
$$\text{Balance owing} = 15\,200$$
$$\text{Total instalments} = 304 \times 60$$
$$= 18\,240$$
$$\text{Total interest} = 18\,240 - 15\,200$$
$$= 3040$$
$$I = Prn$$
$$3040 = 15\,200 \times r \times 5$$
$$3040 = 76\,000r$$
$$r = \frac{3040}{76\,000}$$
$$= 0.04$$

∴ Alyssa is charged interest of 4% p.a.

5. Option B repayments are $260.30 less

$$\text{Deposit} = 0.05 \times 36\,000$$
$$= 1800$$
$$\text{Balance owing} = 34\,200$$

Option A:

$$\text{Interest} = 34\,200 \times 0.04 \times 3$$
$$= 4104$$
$$\text{Total repayments} = 38\,304$$
$$\text{Monthly repayments} = 38\,304 \div 36$$
$$= 1064$$

Option B:

$$\text{Interest} = 34\,200 \times 0.032 \times 4$$
$$= 4377.6$$
$$\text{Total repayments} = 38\,577.6$$
$$\text{Monthly repayments} = 38\,577.6 \div 48$$
$$= 803.7$$
$$\text{Difference} = 1064 - 803.7$$
$$= 260.3$$

∴ Option B repayments are $260.30 less than Option A repayments.

6. 48

$$\text{Deposit} = 0.1 \times 40\,000 = 4000$$
$$\text{Balance owing} = 36\,000$$

Let the number of monthly repayments be x.

∴ the number of years is $\frac{x}{12}$.

$$\text{Interest} = 36\,000 \times 0.06 \times \frac{x}{12}$$
$$\text{Total repayments} = 36\,000 + 36\,000 \times 0.06 \times \frac{x}{12} \ldots ①$$

But, monthly repayments = 930

Total repayments = $930x$ … ②

Let ① = ②

$$36\,000 + 36\,000 \times 0.06 \times \frac{x}{12} = 930x$$
$$36\,000 + 180x = 930x$$
$$750x = 36\,000$$
$$x = \frac{36\,000}{750}$$
$$= 48$$

∴ Jenny made 48 repayments.

Revision Test  (page 22)

1. $24.17

Let her hourly rate be x.

$$1160.16 = 24 \times x + 8 \times 1.5 \times x + 6 \times 2 \times x$$
$$= 24x + 12x + 12x$$
$$= 48x$$
$$x = \frac{1160.16}{48}$$
$$= 24.17$$

∴ Anne's hourly rate is $24.17.

2. $945.50

Total fortnightly tax paid

$= 625 \times 26$

$= 16\,250$

$$\text{Taxable income} = 74\,940 - 1840$$
$$= 73\,100$$

Tax payable

$= 3572 + 0.325 \times 36\,100$

$= 15\,304.5$

$$\text{Refund} = 16\,250 - 15\,304.5$$
$$= 945.5$$

∴ Rod will receive a refund of $945.50.

3. $106 000

Lara's income was between $80 000 and $180 000.

Let her income be x.

$$27\,167 = 17\,547 + 0.37 \times (x - 80\,000)$$
$$9620 = 0.37(x - 80\,000)$$
$$x - 80\,000 = \frac{9620}{0.37}$$
$$= 26\,000$$
$$x = 106\,000$$

∴ Lara had a taxable income of $106 000.

4. \$3600

Subs. $P = 20\,000$, $r = 0.005$, $n = 36$

$$I = Prn$$
$$= 20\,000 \times 0.005 \times 36$$
$$= 3600$$

$\therefore$ simple interest of \$3600 will be earned.

5. 3 years

Subs. $I = 1152$, $P = 9600$, $r = 0.01$

$$I = Prn$$
$$1152 = 9600 \times 0.01 \times n$$
$$1152 = 96n$$
$$n = \frac{1152}{96}$$
$$= 12$$

$$\text{Number of years} = 12 \div 4$$
$$= 3$$

$\therefore$ it will take Liam 3 years.

6. 7% p.a.

$$\text{Deposit} = 0.15 \times 15\,600$$
$$= 2340$$
$$\text{Balance owing} = 13\,260$$
$$\text{Total instalments} = 353.6 \times 48$$
$$= 16\,972.8$$
$$\text{Total interest} = 16\,972.8 - 13\,260$$
$$= 3712.8$$
$$I = Prn$$
$$3712.8 = 13\,260 \times r \times 4$$
$$3712.8 = 53\,040r$$
$$r = \frac{3712.8}{53\,040}$$
$$= 0.07$$

$\therefore$ the interest rate is 7% pa.

Revision Test  (page 23)

1. \$901.32

Let his hourly rate be x.

$$794.02 = 28 \times x + 6 \times 1.5 \times x$$
$$= 28x + 9x$$
$$= 37x$$
$$x = \frac{794.02}{37}$$
$$= 21.46$$

New pay

$= 21.46 \times 30 + 21.46 \times 2 \times 6$

$= 901.32$

$\therefore$ Nathan's pay this week will be \$901.32.

2. \$47.59

$$\text{Annual income} = 2350 \times 26$$
$$= 61\,100$$

Tax payable

$= 3572 + 0.325 \times 24\,100$

$= 11\,404.5$

$$\text{Increased income} = 61\,100 \times 1.03$$
$$= 62\,933$$

New tax payable

$= 3572 + 0.325 \times 25\,933$

$= 12\,000.225$

$$\text{Extra tax} = 12\,000.23 - 11\,404.5$$
$$= 595.73$$

Extra fortnightly tax

$= 22.9126$

$= 22.91$ (2 dec. pl.)

$$\text{Extra weekly income} = 2350 \times 0.03$$
$$= 70.5$$
$$\text{Extra in pay packet} = 70.5 - 22.91$$
$$= 47.59$$

$\therefore$ Nicole will receive an extra \$47.59.

3. \$57 786.15

Anita's tax payable

$= 17\,547 + 0.37 \times (88\,400 - 80\,000)$

$= 20\,655$

As $20\,655 \div 2 = 10\,327.5$, Kerri paid \$10 327.50 tax.

Kerri's income was between \$37 000 and \$80 000.

Let her income be x.

$$10\,327.5 = 3572 + 0.325 \times (x - 37\,000)$$
$$6755.5 = 0.325(x - 37\,000)$$
$$x - 37\,000 = \frac{6755.5}{0.325}$$
$$= 20\,786.15385\ldots$$
$$= 20\,786.15 \text{ (2 dec pl)}$$
$$x = 57\,786.15$$

$\therefore$ Kerri had a taxable income of \$57 786.15.

4. \$740.53

Subs. $P = 5000$, $r = 0.0345$, $n = 2$

$$I = Prn$$
$$= 5000 \times 0.0345 \times 2$$
$$= 345$$
$$\text{Total} = 5000 + 345$$
$$= 5345$$

Subs. $P = 5345$, $r = 0.037$, $n = 2$:

$$I = Prn$$
$$= 5345 \times 0.037 \times 2$$
$$= 395.53$$
$$\text{Total} = 345 + 395.53$$
$$= 740.53$$

$\therefore$ total interest of \$740.53 was earned.

5. 6.5% p.a. and 7.25% p.a.

Let rate on first account be x% p.a.

Let rate on second account be $(x + 0.75)$% p.a.

$$12\,000 \times \frac{x}{100} \times 3$$
$$= 600 + 8000 \times \frac{x + 0.75}{100} \times 3$$
$$360x = 600 + 240(x + 0.75)$$
$$360x = 600 + 240x + 180$$
$$120x = 780$$
$$x = \frac{780}{120}$$
$$= 6.5$$

$6.5 + 0.75 = 7.25$

$\therefore$ interest rates are 6.5% p.a. and 7.25% p.a.

6. 60

$$\text{Deposit} = 0.2 \times 60\,000$$
$$= 12\,000$$
$$\text{Balance owing} = 48\,000$$

Let the number of monthly repayments be x.

$\therefore$ the number of years is $\frac{x}{12}$.

$$\text{Interest} = 48\,000 \times 0.04 \times \frac{x}{12}$$
$$\text{Total repayments} = 48\,000 + 48\,000 \times 0.04 \times \frac{x}{12} \ldots ①$$

But, monthly repayments = 960

Total repayments = $960x$ … ②

Let ① = ②

$$48\,000 + 48\,000 \times 0.04 \times \frac{x}{12} = 960x$$
$$48\,000 + 160x = 960x$$
$$800x = 48\,000$$
$$x = \frac{48\,000}{800}$$
$$= 60$$

$\therefore$ Bruce made 60 repayments.

Key Skill  (pages 24–25)

1. \$3613.50

Subs. $P = 7000$, $r = 0.0425$, $n = 10$

$$A = P(1 + r)^n$$
$$A = 7000(1.0425)^{10}$$
$$= 10\,613.501\,28\ldots$$
$$= 10\,613.50 \text{ (2 dec. pl.)}$$
$$\text{Interest} = 10\,613.50 - 7000$$
$$= 3613.50$$

$\therefore$ Takumi earned interest of \$3613.50.

2. \$2086.03

Monthly: $4.8 \div 12 = 0.4$

$\therefore$ 0.4% per month, or 0.004

Subs. $P = 28\,000$, $r = 0.004$, $n = 18$

$$A = P(1 + r)^n$$
$$A = 28\,000(1.004)^{18}$$
$$= 30\,086.028\,45\ldots$$
$$= 30\,086.03 \text{ (2 dec. pl.)}$$
$$\text{Interest} = 30\,086.03 - 28\,000$$
$$= 2086.03$$

$\therefore$ Ingrid earned interest of \$2086.03.

3. \$522 007

Subs. $P = 462\,000$, $r = 0.031$, $n = 4$

$$A = P(1 + r)^n$$
$$A = 462\,000(1.031)^4$$
$$= 522\,007.3724\ldots$$
$$= 522\,007 \text{ (nearest whole)}$$

$\therefore$ apartment was valued at \$522 007.

4. First account by \$308

1st account:

Subs. $P = 10\,000$, $r = 0.05$, $n = 3$

$$A = P(1 + r)^n$$
$$A = 10\,000(1.05)^3$$
$$= 11\,576.25$$
$$\text{Interest} = 11\,576.25 - 10\,000$$
$$= 1576.25$$

2nd account:

4% p.a. = 1% per quarter;
3 years = 12 quarters

Subs. $P = 10\,000$, $r = 0.01$, $n = 12$

$$A = P(1 + r)^n$$
$$A = 10\,000(1.01)^{12}$$
$$= 11\,268.2503$$
$$= 11\,268.25 \text{ (2 dec pl)}$$
$$\text{Interest} = 11\,268.25 - 10\,000$$
$$= 1268.25$$
$$\text{Difference} = 1576.25 - 1268.25$$
$$= 308$$

∴ first account earns \$308 more interest.

5. 25 166

$$\text{Difference in rates} = 4.2 - 1.8$$
$$= 2.4$$

Subs. $P = 24\,000$, $r = 0.024$, $n = 2$

$$A = P(1 + r)^n$$
$$A = 24\,000(1.024)^2$$
$$= 25\,165.824$$
$$= 25\,166 \text{ (nearest whole)}$$

∴ population is 25 166.

6. \$1479.25

Loan from Lukas:

Subs. $P = 5000$, $r = 0.06$, $n = 8$

$$I = Prn$$
$$= 5000 \times 0.06 \times 8$$
$$= 2400$$

∴ simple interest of \$2400.

Loan to Kate:

Monthly: $7.2 \div 12 = 0.6$

∴ 0.6%/month, or 0.006;
$8 \times 12 = 96$ months

Subs. $P = 5000$, $r = 0.006$, $n = 96$

$$A = P(1 + r)^n$$
$$A = 5000(1.006)^{96}$$
$$= 8879.247332\ldots$$
$$= 8879.25 \text{ (2 dec. pl.)}$$
$$\text{Interest} = 8879.25 - 5000$$
$$= 3879.25$$
$$\text{Difference} = 3879.25 - 2400$$
$$= 1479.25$$

∴ Oscar made \$1479.25.

Key Skill 8 (pages 26–27)

1. \$9400

Subs. $A = 13\,027$, $r = 0.085$, $n = 4$

$$A = P(1 + r)^n$$
$$13\,027 = P(1.085)^4$$
$$P = \frac{13\,027}{1.085^4}$$
$$= 9399.948\,201\ldots$$
$$= 9400 \text{ (nearest whole)}$$

∴ Penelope originally invested \$9400.

2. 6.25%

$26\,000 + 21\,672 = 47\,672$

Subs. $A = 47\,672$, $P = 26\,000$, $n = 10$

$$A = P(1 + r)^n$$
$$47\,672 = 26\,000(1 + r)^{10}$$
$$(1 + r)^{10} = 1.833\,538\,462\ldots$$
$$= 1.83 \text{ (2 dec. pl.)}$$
$$1 + r = \sqrt[10]{1.83}$$
$$r = \sqrt[10]{1.83} - 1$$
$$= 0.062\,500\,155\ldots$$
$$= 0.0625 \text{ (4 dec. pl.)}$$

∴ Emily's investment earned 6.25% p.a.

3. \$2592

Subs. $A = 3000$, $r = 0.05$, $n = 3$

$$A = P(1 + r)^n$$
$$3000 = P(1.05)^3$$
$$P = \frac{3000}{1.05^3}$$
$$= 2591.512\,796\ldots$$
$$= 2592 \text{ (nearest whole)}$$

∴ Ray originally borrowed \$2592.

4. 5%

$9500 + 1527 = 11\,027$

3 years = 12 quarters

Subs. $A = 11\,027$, $P = 9500$, $n = 12$

$$A = P(1 + r)^n$$
$$11\,027 = 9500(1 + r)^{12}$$
$$(1 + r)^{12} = 1.160\,736\,842\ldots$$
$$= 1.16 \text{ (2 dec. pl.)}$$
$$1 + r = \sqrt[12]{1.16}$$
$$r = \sqrt[12]{1.16} - 1$$
$$= 0.012\,498\,715\ldots$$
$$= 0.0125 \text{ (4 dec. pl.)}$$

Interest rate = 1.25%/quarter, or 5% per year

∴ Julia's investment earned 5% p.a.

5. 27 447, 37 000

Subs. $A = 28\,000$, $r = 0.004$, $n = 5$

$$A = P(1 + r)^n$$
$$28\,000 = P(1.004)^5$$
$$P = \frac{28\,000}{1.004^5}$$
$$= 27\,446.657\,78\ldots$$
$$= 27\,447 \text{ (nearest whole)}$$

∴ the population was 27 447 in June 2014.

Subs. $P = 28\,000$, $r = 0.004$, $n = 72$

$$A = P(1 + r)^n$$
$$= 28\,000(1.004)^{72}$$
$$= 37\,323.757\,03$$
$$= 37\,000 \text{ (nearest thousand)}$$

∴ the population will be 37 000 in November 2020.

6. \$4960

Martina:

Subs. $P = 5000$, $r = 0.03$, $n = 5$

$$I = Prn$$
$$= 5000 \times 0.03 \times 5$$
$$= 750$$

∴ simple interest of \$750.

Elena:

$5000 + 750 = 5750$

Subs. $A = 5750$, $r = 0.03$, $n = 5$

$$A = P(1 + r)^n$$
$$5750 = P(1.03)^5$$
$$P = \frac{5750}{1.03^5}$$
$$= 4960.000\,051\ldots$$
$$= 4960 \text{ (nearest whole)}$$

∴ Elena invested \$4960.

Key Skill  (pages 28–29)

1. 8 years

Subs. $A = 22\,738$, $P = 14\,000$, $r = 0.0625$

$$A = P(1 + r)^n$$
$$22\,738 = 14\,000(1.0625)^n$$
$$1.0625^n = \frac{22\,738}{14\,000}$$
$$1.0625^n = 1.624\,142\,857\ldots$$

Try $n = 7$: ∴ $1.0625^7 = 1.528\,63\ldots$

Try $n = 8$: ∴ $1.0625^8 = 1.624\,17\ldots$

∴ $n = 8$ (nearest whole)

∴ Ashley invested for 8 years.

2. 4 years 2 months

Subs. $A = 1000$, $P = 740$, $r = 0.006$

$$A = P(1 + r)^n$$
$$1000 = 740(1.006)^n$$
$$1.006^n = \frac{1000}{740}$$
$$1.006^n = 1.351\,35\ldots$$

Try $n = 30$: ∴ $1.006^{30} = 1.196\,57\ldots$

Try $n = 50$: ∴ $1.006^{50} = 1.348\,64\ldots$

∴ $n = 50$

50 months = 4 years 2 months

∴ after 4 years 2 months the population will be 1000.

3. 35 years

Subs. $A = 210\,000$, $P = 105\,000$, $r = 0.02$

$A = P(1 + r)^n$

$210\,000 = 105\,000(1.02)^n$

$1.02^n = 2$

Try $n = 30$: $\therefore 1.02^{30} = 1.811\,36\ldots$

Try $n = 35$: $\therefore 1.02^{35} = 1.999\,88\ldots$

$\therefore n = 35$ (nearest whole)

$\therefore$ it will double in 35 years.

4. August

Subs. $A = 1000$, $P = 16$, $r = 0.06$

$A = P(1 + r)^n$

$1000 = 16(1.06)^n$

$1.06^n = \frac{1000}{16}$

$1.06^n = 62.5$

Try $n = 60$: $\therefore 1.06^{60} = 32.98\ldots$

Try $n = 70$: $\therefore 1.06^{70} = 59.075\ldots$

Try $n = 71$: $\therefore 1.06^{71} = 62.6204\ldots$

$\therefore n = 71$ (nearest whole)

$\therefore$ 71 days is 2 months 10 days.

$\therefore$ in August there will be 1000 cases.

5. 63

Subs. $A = 100\,000$, $P = 5000$, $r = 0.048$

$A = P(1 + r)^n$

$100\,000 = 5000(1.048)^n$

$1.048^n = \frac{100\,000}{5000}$

$1.048^n = 20$

Try $n = 60$: $\therefore 1.048^{60} = 16.660\ldots$

Try $n = 63$: $\therefore 1.048^{63} = 19.1761\ldots$

Try $n = 64$: $\therefore 1.048^{64} = 20.0965\ldots$

The investment will be worth $100 000 just before her 64th birthday.

$\therefore$ Katia will be 63.

6. 5 months

Abigail:

Subs. $I = 5000$, $P = 20\,000$, $r = 0.005$

$I = Prn$

$5000 = 20\,000 \times 0.005 \times n$

$5000 = 100n$

$n = \frac{5000}{100} = 50$

$\therefore$ it will take Abigail 50 months.

Keith:

$20\,000 + 5000 = 25\,000$

Subs. $A = 25\,000$, $P = 20\,000$, $r = 0.005$

$A = P(1 + r)^n$

$25\,000 = 20\,000(1.005)^n$

$1.005^n = 1.25$

Try $n = 35$: $\therefore 1.005^{35} = 1.1907\ldots$

Try $n = 40$: $\therefore 1.005^{40} = 1.2207\ldots$

Try $n = 45$: $\therefore 1.005^{45} = 1.2516\ldots$

$\therefore n = 45$ (nearest whole)

Difference $= 50 - 45 = 5$

$\therefore$ Keith will earn the interest 5 months earlier than Abigail.

Key Skill  10 (pages 30–31)

1. $38 075.75

Subs. $P = 62\,000$, $r = 0.15$, $n = 3$

$A = P(1 - r)^n$

$= 62\,000(0.85)^3$

$= 38\,075.75$

$\therefore$ the caravan was valued at $38 075.75.

2. $1300

Subs. $A = 5.24$, $r = 0.015$, $n = 10$

$A = P(1 - r)^n$

$5.24 = P(0.985)^{10}$

$P = \frac{5.24}{0.985^{10}}$

$= 6.094\,933\,647\ldots$

$= 6.09$ (2 dec. pl.)

$\therefore$ the price was $6.09 per share.

Loss per share $= 6.09 - 5.24$

$= 0.85$

Sergie's total loss $= 0.85 \times 1500$

$= 1275$

$= 1300$ (nearest 100)

$\therefore$ Sergie lost $1300.

3. $4200

Subs. $P = 82\,000$, $r = 0.18$, $n = 15$

$A = P(1 - r)^n$

$= 82\,000(0.82)^{15}$

$= 4178.511\,85\ldots$

$= 4200$ (nearest hundred)

$\therefore$ the tractor will be worth $4200.

4. 16%

Subs. $A = 45\,804$, $P = 92\,000$, $n = 4$

$A = P(1 - r)^n$

$45\,804 = 92\,000(1 - r)^4$

$(1 - r)^4 = \frac{45\,804}{92\,000}$

$1 - r = \sqrt[4]{\frac{45\,804}{92\,000}}$

$= 0.839\,999\,243\ldots$

$= 0.84$ (2 dec. pl.)

$r = 1 - 0.84$

$= 0.16$

$\therefore$ the motorhome depreciates at 16% p.a.

5. 1280 euros

Subs. $P = 0.68$, $r = 0.012$, $n = 5$

$A = P(1 - r)^n$

$= 0.68(0.988)^5$

$= 0.640\,167\,519\ldots$

$= 0.6402$ (4 dec. pl.)

Number of euros $= 2000 \times 0.6402$

$= 1280.4$

$= 1280$ (nearest ten)

$\therefore$ $2000 can be exchanged for 1280 euros.

6. $279 000

Subs. $P = 240\,000$, $r = 0.07$, $n = 6$

$A = P(1 + r)^n$

$= 240\,000(1.07)^6$

$= 360\,175.2844\ldots$

$= 360\,175.28$ (2 dec. pl.)

Subs. $P = 360\,175.28$, $r = 0.12$, $n = 2$

$A = P(1 - r)^n$

$= 360\,175.28(0.88)^2$

$= 278\,919.7403\ldots$

$= 279\,000$ (nearest thousand)

$\therefore$ the unit was valued at $279 000.

Revision Test  3 (page 32)

1. $1581.94

Subs. $P = 6400$, $r = 0.0375$, $n = 6$

$A = P(1 + r)^n$

$A = 6400(1.0375)^6$

$= 7981.942\,709\ldots$

$= 7981.94$ (2 dec. pl.)

Interest $= 7981.94 - 6400$

$= 1581.94$

$\therefore$ Ruby earned interest of $1581.94.

2. $1 148 000

Subs. $P = 820\,000$, $r = 0.043$, $n = 8$

$A = P(1 + r)^n$

$A = 820\,000(1.043)^8$

$= 1\,148\,387.138\ldots$

$= 1\,148\,000$ (nearest thousand)

$\therefore$ the price of the painting was $1 148 000.

3. 4.5%

$14\,400 + 6078 = 20\,478$

Subs. $A = 20\,478$, $P = 14\,400$, $n = 8$

$A = P(1 + r)^n$

$20\,478 = 14\,400(1 + r)^8$

$(1 + r)^8 = 1.422\,083\,333\ldots$

$= 1.42$ (2 dec. pl.)

$1 + r = \sqrt[8]{1.42}$

$r = \sqrt[8]{1.42} - 1$

$= 0.044\,998\,412\ldots$

$= 0.045$ (3 dec. pl.)

$\therefore$ Bethany's investment earned 4.5% p.a.

4. 18 years

Subs. $A = 32\,800$, $P = 16\,400$, $r = 0.04$

$A = P(1 + r)^n$

$32\,800 = 16\,400(1.04)^n$

$1.04^n = 2$

Try $n = 15$: $\therefore 1.04^{15} = 1.800\,94\ldots$

Try $n = 17$: $\therefore 1.04^{17} = 1.947\,90\ldots$

Try $n = 18$: $\therefore 1.04^{18} = 2.025\,81\ldots$

$\therefore n = 18$ (nearest whole)

$\therefore$ its value will double in 18 years.

5. $6200

Subs. $P = 38\,000$, $r = 0.26$, $n = 6$

$$\begin{aligned} A &= P(1 - r)^n \\ &= 38\,000(0.74)^6 \\ &= 6239.846\,627\ldots \\ &= 6200 \text{ (nearest hundred)} \end{aligned}$$

$\therefore$ the vehicle will be worth \$6200.

6. 8%

Subs. $A = 51\,175.80$, $P = 78\,000$, $n = 5$

$$\begin{aligned} A &= P(1 - r)^n \\ 51\,175.8 &= 78\,000(1 - r)^5 \\ (1 - r)^5 &= \frac{51\,175.8}{78\,000} \\ 1 - r &= \sqrt[5]{\frac{51\,175.8}{78\,000}} \\ &= 0.919\,166\,118\ldots \\ &= 0.92 \text{ (2 dec. pl.)} \\ r &= 1 - 0.92 \\ &= 0.08 \end{aligned}$$

$\therefore$ the caravan depreciated at 8% p.a.

Revision Test  (page 33)

1. 24 100, 31 900

Subs. $A = 25\,200$, $r = 0.0036$, $n = 12$

$$\begin{aligned} A &= P(1 + r)^n \\ 25\,200 &= P(1.0036)^{12} \\ P &= \frac{25\,200}{(1.0036)^{12}} \\ &= 24\,136.411\,92\ldots \\ &= 24\,100 \text{ (nearest hundred)} \end{aligned}$$

5 years 6 months = 66 months

Subs. $P = 25\,200$, $r = 0.0036$, $n = 66$

$$\begin{aligned} A &= P(1 + r)^n \\ &= 25\,200 \times (1.0036)^{66} \\ &= 31\,945.053\,16\ldots \\ &= 31\,900 \text{ (nearest hundred)} \end{aligned}$$

$\therefore$ population was 24 100 in April 2012, and estimated to be 31 900 in October 2018.

2. 1.6%

$14\,000 + 1164 = 15\,164$

Subs. $A = 15\,164$, $P = 14\,000$, $n = 20$

$$\begin{aligned} A &= P(1 + r)^n \\ 15\,164 &= 14\,000(1 + r)^{20} \\ (1 + r)^{20} &= 1.083\,142\,857\ldots \\ &= 1.08 \text{ (2 dec. pl.)} \\ 1 + r &= \sqrt[20]{1.08} \\ r &= \sqrt[20]{1.08} - 1 \\ &= 0.004\,001\,327\ldots \\ &= 0.004 \text{ (3 dec. pl.)} \end{aligned}$$

$\therefore$ earned 0.4% per quarter.

$\therefore$ Natalia's investment earned 1.6% p.a.

3. $4710.10

Loan from Adam:

Subs. $P = 16\,000$, $r = 0.05$, $n = 6$

$$\begin{aligned} I &= Prn \\ &= 16\,000 \times 0.05 \times 6 \\ &= 4800 \end{aligned}$$

$\therefore$ simple interest of \$4800.

Loan to Magda:

Monthly: $7.8 \div 12 = 0.65$

$\therefore$ 0.65%/month, or 0.0065;

$6 \times 12 = 72$ mths

Subs. $P = 16\,000$, $r = 0.0065$, $n = 72$

$$\begin{aligned} A &= P(1 + r)^n \\ A &= 16\,000(1.0065)^{72} \\ &= 25\,510.095\,64\ldots \\ &= 25\,510.10 \text{ (2 dec pl)} \\ \text{Interest} &= 25\,510.10 - 16\,000 \\ &= 9510.10 \end{aligned}$$

$\therefore$ compound interest of \$9510.10.

$$\begin{aligned} \text{Difference} &= 9510.10 - 4800 \\ &= 4710.10 \end{aligned}$$

$\therefore$ Gemma made \$4710.10.

4. $3944

Emma:

Subs. $A = 14\,000$, $P = 10\,000$, $r = 0.005$

$$\begin{aligned} A &= P(1 + r)^n \\ 14\,000 &= 10\,000(1.005)^n \\ 1.005^n &= \frac{14\,000}{10\,000} \\ &= 1.4 \end{aligned}$$

Try $n = 60$: $\therefore 1.005^{60} = 1.3488\ldots$

Try $n = 67$: $\therefore 1.005^{67} = 1.3967\ldots$

$\therefore$ Emma invested for 67 months.

Gavin:

Subs. $P = 8000$, $r = 0.006$, $n = 67$

$$\begin{aligned} A &= P(1 + r)^n \\ &= 8000(1.006)^{67} \\ &= 11\,944.134\,77\ldots \\ &= 11\,944 \text{ (nearest whole)} \\ \text{Interest} &= 11\,944 - 8000 \\ &= 3944 \end{aligned}$$

$\therefore$ Gavin has earned \$3944.

5. $6261

Subs. $P = 32\,000$, $r = 0.18$, $n = 5$

$$\begin{aligned} A &= P(1 - r)^n \\ &= 32\,000(0.82)^5 \\ &= 11\,863.674\,98\ldots \\ &= 11\,863.67 \text{ (2 dec. pl.)} \end{aligned}$$

Subs. $P = 11\,863.67$, $r = 0.12$, $n = 5$

$$\begin{aligned} A &= P(1 - r)^n \\ &= 11\,863.67(0.88)^5 \\ &= 6260.837\,309\ldots \\ &= 6261 \text{ (nearest whole)} \end{aligned}$$

$\therefore$ the car is valued at \$6261.

6. $357 000

Subs. $P = 310\,000$, $r = 0.046$, $n = 8$

$$\begin{aligned} A &= P(1 + r)^n \\ &= 310\,000(1.046)^8 \\ &= 444\,237.4526\ldots \\ &= 444\,237.45 \text{ (2 dec. pl.)} \end{aligned}$$

Subs. $P = 444\,237.45$, $r = 0.053$, $n = 4$

$$\begin{aligned} A &= P(1 - r)^n \\ &= 444\,237.45(0.947)^4 \\ &= 357\,285.2469\ldots \\ &= 357\,000 \text{ (nearest thousand)} \end{aligned}$$

$\therefore$ the unit is valued at \$357 000.

Key Skill  (pages 34–35)

1. $3x + 4$

Lincoln's age 3 years ago $= x - 3$

'twice as old': $2(x - 3)$

'In 5 years time': $x + 5$ and $2(x - 3) + 5$

$$\begin{aligned} \text{Sum} &= x + 5 + 2(x - 3) + 5 \\ &= x + 5 + 2x - 6 + 5 \\ &= 3x + 4 \end{aligned}$$

$\therefore$ the sum is $3x + 4$.

2. $\$(\frac{x}{20} + 2)$

There were $(40 - x)$ 5-cent coins.

Total value

$= 10 \times x + 5 \times (40 - x)$

$= 10x + 200 - 5x$

$= 5x + 200$

$$\begin{aligned} \therefore (5x + 200) \text{ cents} &= \$(\frac{5x + 200}{100}) \\ &= \$(\frac{x}{20} + 2) \end{aligned}$$

$\therefore$ Oliver had $\$(\frac{x}{20} + 2)$ in his money jar.

3. $\$(\frac{x(p + 2q)}{p + q})$

$$\begin{aligned} \text{Total cost} &= p \times x + q \times 2x \\ &= px + 2qx \\ \text{Total bags} &= p + q \\ \text{Average cost} &= \frac{px + 2qx}{p + q} \\ &= \frac{x(p + 2q)}{p + q} \end{aligned}$$

$\therefore$ the average cost was $\$(\frac{x(p + 2q)}{p + q})$ per bag.

4. $(5x + 1000)

There were $(200 - x)$ children's tickets.

Price of child's ticket = \$5.

$$\begin{aligned} \text{Total value} &= 10x + 5(200 - x) \\ &= 10x + 1000 - 5x \\ &= 5x + 1000 \end{aligned}$$

$\therefore$ the total value of ticket sales was $\$(5x + 1000)$.

5. $\mathbf{\dfrac{r(m + n) - (mp + nq)}{m + n}}$

Total cost $= m \times p + n \times q$

$= mp + nq$

Total selling price $= r(m + n)$

Total profit $= r(m + n) - mp + nq$

Total boxes $= m + n$

Average profit

$= \dfrac{r(m + n) - (mp + nq)}{m + n}$

6. $\mathbf{(\dfrac{x + 4y + 720}{240})}$ **hours**

Time for first journey $= \dfrac{x}{60}$

Time of stoppage (in h) $= \dfrac{y}{60}$

Time for second journey $= \dfrac{240 - x}{80}$

Total time $= \dfrac{x}{60} + \dfrac{y}{60} + \dfrac{240 - x}{80}$

$= \dfrac{4x + 4y + 3(240 - x)}{240}$

$= \dfrac{4x + 4y + 720 - 3x}{240}$

$= \dfrac{x + 4y + 720}{240}$

∴ it took Ruby $\dfrac{x + 4y + 720}{240}$ hours.

Key Skill  12 (pages 36–37)

1. $\mathbf{\dfrac{7x - 5}{7x}}$

Craig worked $3x$ h at normal pay and $2x$ h at double time.

Total pay $= 3x \times y + 2x \times 2 \times y$

$= 7xy$

Amount saved $= 7xy - 5y$

Fraction saved $= \dfrac{7xy - 5y}{7xy}$

$= \dfrac{y(7x - 5)}{7xy}$

$= \dfrac{7x - 5}{7x}$

∴ Craig saved $\dfrac{7x - 5}{7x}$ of his total pay.

2. $\mathbf{\$(\dfrac{x + 80\,000}{50})}$

Gaby has $\$x$ earning 5% p.a. interest and $\$(20\,000 - x)$ earning 4% p.a. interest.

Use $I = Prn$.

Total interest

$= x \times \dfrac{5}{100} \times 2$

$+ (20\,000 - x) \times \dfrac{4}{100} \times 2$

$= \dfrac{10x}{100} + \dfrac{8}{100}(20\,000 - x)$

$= \dfrac{10x}{100} + \dfrac{160\,000 - 8x}{100}$

$= \dfrac{2x + 160\,000}{100}$

$= \dfrac{2(x + 80\,000)}{100}$

$= \dfrac{x + 80\,000}{50}$

∴ Gaby earned interest of $\$(\dfrac{x + 80\,000}{50})$.

3. $\mathbf{\dfrac{xy + (30 - x)(y + 1)}{30}}$

There are x students who are y years old.

There are $(30 - x)$ students who are $(y + 1)$ years old.

Average age

$= \dfrac{xy + (30 - x)(y + 1)}{30}$

$= \dfrac{xy + 30y + 30 - xy - x}{30}$

$= \dfrac{30y - x + 30}{30}$

∴ the average age is $\dfrac{xy + (30 - x)(y + 1)}{30}$ (or $\dfrac{30y - x + 30}{30}$).

4. $\mathbf{\$(\dfrac{ac(b - a)}{100})}$

Length + breadth $= b$

breadth $= b - a$

Area $= a(b - a)$

Cost in cents $= c \times a(b - a)$

$= ac(b - a)$

Cost in dollars $= \dfrac{ac(b - a)}{100}$

∴ the cost is $\$(\dfrac{ac(b - a)}{100})$.

5. $\mathbf{\$(\dfrac{5mx}{2}(1 + \dfrac{p}{100}))}$

New hourly rate $= x + \dfrac{p}{100} \times x$

$= x(1 + \dfrac{p}{100})$

Total pay

$= x(1 + \dfrac{p}{100}) \times (m + 1\dfrac{1}{2}m)$

$= x(1 + \dfrac{p}{100}) \times \dfrac{5m}{2}$

$= \dfrac{5mx}{2}(1 + \dfrac{p}{100})$

∴ Ginnifer will earn $\$(\dfrac{5mx}{2}(1 + \dfrac{p}{100}))$.

6. $\mathbf{\$(x - 12\,000)(1 + \dfrac{y}{1200})^{12}}$

Amount invested $= x - 12\,000$

Subs. $P = x - 12\,000$, $r = \dfrac{y}{1200}$, $n = 12$

$A = P(1 + r)^n$

$= (x - 12\,000)(1 + \dfrac{y}{1200})^{12}$

∴ Josh will have $\$(x - 12\,000)(1 + \dfrac{y}{1200})^{12}$

Key Skill  13 (pages 38–39)

1. 39 and 57

Let one stall sell x muffins.

∴ the other stall sells $x + 18$.

$x + x + 18 = 96$

$2x + 18 = 96$

$2x = 96 - 18$

$2x = 78$

$x = 39$

Now, $x + 18 = 39 + 18$

$= 57$

∴ the stalls sold 39 and 57 muffins.

2. 140 cm, 60 cm

Let the length of the long piece be x cm.

∴ the length of the short piece is $(x - 80)$ cm.

$x + x - 80 = 200$

$2x - 80 = 200$

$2x = 200 + 80$

$2x = 280$

$x = 140$

Now, $x - 80 = 140 - 80$

$= 60$

∴ the lengths are 140 cm and 60 cm.

3. 1.9 kg

Let the front-yard amount be x kg.

∴ the backyard amount is $(x + 1.2)$ kg.

$x + x + 1.2 = 5$

$2x + 1.2 = 5$

$2x = 5 - 1.2$

$2x = 3.8$

$x = 1.9$

∴ 1.9 kg of fertiliser was used on the front lawn.

4. 8 km

Let the 'certain distance' be x km.

∴ the distance run is $(x + 2)$ km and the distance hitched is $(x + 6)$ km.

$x + x + 2 + x + 6 = 14$

$3x + 8 = 14$

$3x = 14 - 8$

$3x = 6$

$x = 2$

Now, $x + 6 = 2 + 6$

$= 8$

∴ Kenny travelled 8 km in the car.

5. 430 km

Let the distance travelled Saturday be x km.

$\therefore$ the distance travelled on Sunday was $(x - 80)$ km.

$$\begin{aligned} x + x - 80 &= 940 \\ 2x - 80 &= 940 \\ 2x &= 940 + 80 \\ 2x &= 1020 \\ x &= 510 \\ \text{Now, } x - 80 &= 510 - 80 \\ &= 430 \end{aligned}$$

$\therefore$ Ehmi drove 430 km on Sunday.

6. \$357

Let the number of double-time hours be x.

$\therefore$ the number of normal hours is $(x - 3)$.

$$\begin{aligned} x + x - 3 &= 16 \\ 2x - 3 &= 16 \\ 2x &= 16 + 3 \\ 2x &= 19 \\ x &= 9.5 \\ \text{Now, } x - 3 &= 9.5 - 3 \\ &= 6.5 \end{aligned}$$

$\therefore$ Fiona worked 6.5 hours normal and 9.5 hours double time.

$$\begin{aligned} \text{Total pay} &= 14 \times 6.5 + 14 \times 2 \times 9.5 \\ &= 357 \end{aligned}$$

$\therefore$ Fiona is paid \$357.

Key Skill  14 (pages 40–41)

1. 8 green, 16 red, 32 blue

Let x be the number of green balls.

$\therefore$ the number of red balls is $2x$ and blue balls is $4x$.

$$\begin{aligned} x + 2x + 4x &= 56 \\ 7x &= 56 \\ x &= 8 \end{aligned}$$

$\therefore$ there are 8 green balls, 16 red balls and 32 blue balls.

2. 66

Let Carmen's age be x.

$\therefore$ Jack's age now is $5x$.

In 12 years, Carmen will be $(x + 12)$, Jack will be $(5x + 12)$.

$$\begin{aligned} x + 12 &= \frac{1}{2}(5x + 12) \\ 2(x + 12) &= 5x + 12 \\ 2x + 24 &= 5x + 12 \\ 5x - 2x &= 24 - 12 \\ 3x &= 12 \\ x &= 4 \end{aligned}$$

Today, Carmen is 4 and Jack is 20.

$\therefore$ the difference in age is 16 years.

$50 + 16 = 66$

$\therefore$ Jack will be 66 years old.

3. 9 and 27

Let the smaller number be x.

$\therefore$ the larger number is $3x$.

$$\begin{aligned} 3x + 12 &= x + 30 \\ 3x - x &= 30 - 12 \\ 2x &= 18 \\ x &= 9 \end{aligned}$$

$\therefore$ the numbers are 9 and 27.

4. Jacob 20, Anthony 16

Today: Let Anthony's age be x.

$\therefore$ Jacob's age is $(x + 4)$.

9 years ago: Anthony was $(x - 9)$ and Christopher was $3(x - 9)$.

$\therefore$ today, Christopher is $3(x - 9) + 9$.

$$\begin{aligned} x + x + 4 + 3(x - 9) + 9 &= 66 \\ x + x + 4 + 3x - 27 + 9 &= 66 \\ 5x - 14 &= 66 \\ 5x &= 66 + 14 \\ 5x &= 80 \\ x &= 16 \end{aligned}$$

$\therefore$ today, Anthony is 16, Jacob is 20 and Christopher is 30.

The party is today!

$\therefore$ Jacob is 20 and Anthony is 16.

5. Xavier: 216 000, Zoe: 432 000, Hannah: 72 000

Let Xavier's point-score be x.

$\therefore$ Zoe's point-score is $2x$ and Hannah's point-score is $\frac{x}{3}$.

$$\begin{aligned} x + 2x + \frac{x}{3} &= 720\,000 \\ 3x + 6x + x &= 2\,160\,000 \\ 10x &= 2\,160\,000 \\ x &= 216\,000 \end{aligned}$$

$\therefore 2x = 432\,000$ and $\frac{x}{3} = 72\,000$.

$\therefore$ Xavier has 216 000, Zoe has 432 000 and Hannah has 72 000.

6. 24

Let the number of males be x.

$\therefore$ the number of females is $(60 - x)$.

$$\begin{aligned} \frac{x}{2} + \frac{60 - x}{3} &= 26 \\ 6 \times \frac{x}{2} + 6 \times \frac{60 - x}{3} &= 6 \times 26 \\ 3x + 2(60 - x) &= 156 \\ 3x + 120 - 2x &= 156 \\ x + 120 &= 156 \\ x &= 156 - 120 \\ x &= 36 \\ \text{Also, } 60 - x &= 60 - 36 \\ &= 24 \end{aligned}$$

$\therefore$ 24 females were at the party.

Key Skill  15 (pages 42–43)

1. 85 m, 60 m and 45 m

$$\begin{aligned} \text{Perimeter} &= 1235 \div 6.5 \\ &= 190 \end{aligned}$$

Let the length of the longest side be x.

Let the length of a second side be $(x - 25)$.

Let the length of a third side be $(x - 40)$.

$$\begin{aligned} x + x - 25 + x - 40 &= 190 \\ 3x - 65 &= 190 \\ 3x &= 255 \\ x &= 85 \end{aligned}$$

$85 - 25 = 60$, $85 - 40 = 45$.

$\therefore$ the sides are 85 m, 60 m and 45 m.

2. 176 cm²

Let the width of the rectangle be x.

Let the length of the rectangle be $(2x + 6)$.

$$\begin{aligned} 2(x + 2x + 6) &= 60 \\ 2(3x + 6) &= 60 \\ 6x + 12 &= 60 \\ 6x &= 60 - 12 \\ 6x &= 48 \\ x &= 8 \end{aligned}$$

$2 \times 8 + 6 = 22$

$\therefore$ dimensions are 22 cm and 8 cm.

$$\begin{aligned} \text{Area} &= 22 \times 8 \\ &= 176 \end{aligned}$$

$\therefore$ area of rectangle is 176 cm^2.

3. 32 cm²

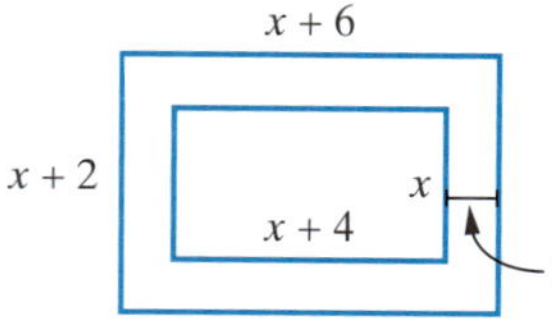

Let the width of the image be x.

$\therefore$ the length of the image is $(x + 4)$.

The dimensions of each photo are $(x + 6)$ and $(x + 2)$.

Area of border:

$$\begin{aligned} (x + 6)(x + 2) - x(x + 4) &= 20 \\ x^2 + 8x + 12 - x^2 - 4x &= 20 \\ 4x + 12 &= 20 \\ 4x &= 8 \\ x &= 2 \end{aligned}$$

$$\begin{aligned} \text{Area of photo} &= (2 + 6)(2 + 2) \\ &= 32 \end{aligned}$$

$\therefore$ area of photo is 32 cm^2.

4. 98 m²

Let the width of the yard be x.

$\therefore$ the length of the yard is $2x$.

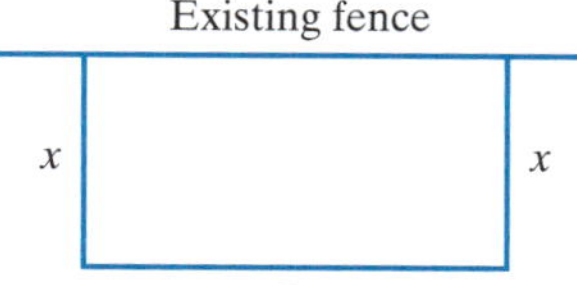

$$2x + x + x = 28$$
$$4x = 28$$
$$x = 7$$

Dimensions are 14 m by 7 m.

$$\text{Area} = 14 \times 7 = 98$$

$\therefore$ area of yard is 98 m².

5. 70 m, 110 m and 180 m

$$\text{Perimeter} = 2970 \div 8.25 = 360$$

Let the length of the one side be x.

Let the length of a second side be $(x - 40)$.

Let the length of a third side be $(x + 70)$.

$$x + x - 40 + x + 70 = 360$$
$$3x + 30 = 360$$
$$3x = 330$$
$$x = 110$$

$110 - 40 = 70$, $110 + 70 = 180$.

$\therefore$ the sides are 70 m, 110 m and 180 m.

6. $24

Let the width be x.

$\therefore$ the length is $(2x - 10)$.

$$\text{Perimeter} = 2(2x - 10 + x) = 2(3x - 10)$$

Also, $64 \div 40 = 1.6$.

1.6 m = 160 cm

$$\therefore 2(3x - 10) = 160$$
$$6x - 20 = 160$$
$$6x = 180$$
$$x = 30$$

Now, $2x - 10 = 2(30) - 10 = 50$

Dimensions are 50 cm by 30 cm, or 0.5 m by 0.3 m.

$$\text{Area} = 0.5 \times 0.3 = 0.15$$
$$\text{Cost of glass} = 0.15 \times 160 = 24$$

$\therefore$ the cost is $24.

Key Skill 16 (pages 44–45)

1. 27° (or 63°)

Let the angle be x.

$\therefore$ the complement is $(90 - x)$ and the supplement is $(180 - x)$.

$$3(90 - x) - (180 - x) = 36$$
$$270 - 3x - 180 + x = 36$$
$$90 - 2x = 36$$
$$2x = 90 - 36$$
$$2x = 54$$
$$x = 27$$

$\therefore$ the angle was 27°.

[Note: 63° is also a correct answer.]

2. 40°, 60°, 80°

Let the smallest angle be x.

$\therefore$ the other angles are $2x$ and $(x + 20)$.

$$x + 2x + x + 20 = 180$$
$$4x + 20 = 180$$
$$4x = 160$$
$$x = 40$$

$2x = 2(40) = 80$

$x + 20 = 40 + 20 = 60$

$\therefore$ the angles are 40°, 60° and 80°.

3. 41°, 139°

Let the smaller angle be x.

$\therefore$ another angle is $(57 + 2x)$.

$$x + 57 + 2x = 180$$
$$3x + 57 = 180$$
$$3x = 123$$
$$x = 41$$

Supplement of 41° is 139°.

$\therefore$ angles are 41° and 139° (and 41° and 139°).

4. 50°, 50°, 80° or 40°, 70°, 70°

Let the smaller angle be x.

$\therefore$ let another angle be $x + 30$.

Case 1: $x + x + x + 30 = 180$
$$3x + 30 = 180$$
$$3x = 150$$
$$x = 50$$

$\therefore$ angles are 50°, 50° and 80°.

Case 2: $x + x + 30 + x + 30 = 180$
$$3x + 60 = 180$$
$$3x = 120$$
$$x = 40$$

$\therefore$ angles are 40°, 70° and 70°.

5. 16°, 74°, 90°

Let the smallest angle be x.

Let another angle be $4x + 10$.

$$x + 4x + 10 + 90 = 180$$
$$5x + 100 = 180$$
$$5x = 80$$
$$x = 16$$

$\therefore$ angles are 16°, 74° and 90°.

6. 8 : 7 : 21

Let the first angle be x.

The complement is $(90 - x)$ and the supplement is $(180 - x)$.

$$180 - x = 40 + 2(90 - x)$$
$$180 - x = 40 + 180 - 2x$$
$$x = 40$$

Let the second angle be $3y$.

$\therefore$ the third angle is y.

$$3y + y + 40 = 180$$
$$4y + 40 = 180$$
$$4y = 140$$
$$y = 35$$

$\therefore$ angles are 40°, 35° and 105°.

$40 : 35 : 105 = 8 : 7 : 21$

Key Skill 17 (pages 46–47)

1. 1 pm

Let Jayden's time of travel be x.

$\therefore$ Mia's time of travel is $x - 1$.

As Distance = Speed × Time:
$$60 \times x = 90(x - 1)$$
$$60x = 90x - 90$$
$$30x = 90$$
$$x = 3$$

10 am plus 3 h = 1 pm

$\therefore$ Mia caught up to Jayden at 1 pm.

2. 105 km

$$\text{Elapsed time} = 8{:}50 - 8{:}30 = 20 \text{ min} = \frac{1}{3}\text{ h}$$

Let Malia's time of journey be x.

$\therefore$ Christian's time of journey is $x - \frac{1}{3}$

As Distance = Speed × Time:
$$70 \times x = 90(x - \frac{1}{3})$$
$$70x = 90x - 30$$
$$20x = 30$$
$$x = 1.5$$

After 1.5 hours, Christian caught up to Malia.

$$\text{Distance travelled} = 70 \times 1.5 = 105$$

$\therefore$ Christian will have to drive 105 km to catch Malia.

3. 10:15 am

Let the motorboat's time of travel be x.

$\therefore$ the cabin cruiser's time of travel is $x - \frac{1}{2}$

As Distance = Speed × Time:
$$12 \times x = 20(x - \frac{1}{2})$$
$$12x = 20x - 10$$
$$8x = 10$$
$$x = 1.25$$

1.25 h = 1 h 15 min

9 am plus 1 h 15 min = 10:15 am

$\therefore$ the cabin cruiser reached the motorboat at 10:15 am.

4. 6 pm, 240 km from Perth

Let Michael's time of travel be x.

$\therefore$ Michelle's time of travel is $x - 1$.

As Distance = Speed × Time:

$$\begin{aligned} 80 \times x + 90(x - 1) &= 420 \\ 80x + 90x - 90 &= 420 \\ 170x &= 420 + 90 \\ 170x &= 510 \\ x &= 3 \end{aligned}$$

3 pm plus 3 h = 6:00 pm

$$\begin{aligned} \text{Michael's distance} &= 80 \times 3 \\ &= 240 \end{aligned}$$

$\therefore$ Michael and Michelle meet at 6:00 pm, 240 km from Perth (or 180 km from Geraldton).

5. 25 min

Let Rhianna's time of travel be x.

$\therefore$ flatmate's time of travel is $x - \frac{1}{12}$.

As Distance = Speed × Time:

$$\begin{aligned} 60 \times x &= 72(x - \tfrac{1}{12}) \\ 60x &= 72x - 6 \\ 12x &= 6 \\ x &= 0.5 \end{aligned}$$

0.5 h = 30 min

8:00 am plus 30 min = 8:30

Flatmate's travel time

= 8:30 minus 8:05

= 25 min

$\therefore$ it took 25 minutes for the flatmate to catch up to Rhianna.

6. 2.5 km, $6\frac{2}{3}$ km/h

$45 \text{ min} = \frac{3}{4} \text{ h}$

Let the distance to school be x.

As Time = Distance ÷ Speed:

$$\begin{aligned} \frac{x}{5} + \frac{x}{10} &= \frac{3}{4} \\ 20 \times \frac{x}{5} + 20 \times \frac{x}{10} &= 20 \times \frac{3}{4} \\ 4x + 2x &= 15 \\ 6x &= 15 \\ x &= 2.5 \end{aligned}$$

$\therefore$ the distance from home to school is 2.5 km (both ways = 5 km).

$$\begin{aligned} \text{Average speed} &= \frac{5}{0.75} \\ &= 6\frac{2}{3} \end{aligned}$$

$\therefore$Mitch's average speed was $6\frac{2}{3}$ km/h.

Key Skill  18 (pages 48–49)

1. 120 childen, 200 adults

Let the number of children's tickets be x.

$\therefore$ number of adult tickets is $320 - x$.

$$\begin{aligned} 8 \times x + 12(320 - x) &= 3360 \\ 8x + 3840 - 12x &= 3360 \\ 3840 - 4x &= 3360 \\ 4x &= 3840 - 3360 \\ 4x &= 480 \\ x &= 120 \end{aligned}$$

Also, $320 - 120 = 200$

$\therefore$ 120 children and 200 adults bought tickets.

2. 5 kg sultanas, 15 kg peanuts

Let the mass of sultanas be x.

$\therefore$ mass of peanuts be $20 - x$.

$$\begin{aligned} 12 \times x + 16(20 - x) &= 15 \times 20 \\ 12x + 320 - 16x &= 300 \\ 320 - 4x &= 300 \\ 4x &= 20 \\ x &= 5 \end{aligned}$$

$$\begin{aligned} \text{Also, } 20 - x &= 20 - 5 \\ &= 15 \end{aligned}$$

$\therefore$ 5 kg of sultanas and 15 kg of peanuts should be used.

3. 6

Let the number of apple pies be x.

$\therefore$ the number of meat pies was $3x$.

$$\begin{aligned} 10 \times x + 8 \times 3x &= 68 \\ 10x + 24x &= 68 \\ 34x &= 68 \\ x &= 2 \end{aligned}$$

$$\begin{aligned} \text{Also, } 3x &= 3 \times 2 \\ &= 6 \end{aligned}$$

$\therefore$ Mary bought 6 meat pies.

4. 1.6 kg of Batch A, 1.4 kg of Batch B

Let the mass of Batch A be x.

$\therefore$ mass of Batch B be $3 - x$.

$$\begin{aligned} 24 \times x + 32(3 - x) &= 83.2 \\ 24x + 96 - 32x &= 83.2 \\ 96 - 8x &= 83.2 \\ 8x &= 12.8 \\ x &= 1.6 \end{aligned}$$

$$\begin{aligned} \text{Also, } 3 - x &= 3 - 1.6 \\ &= 1.4 \end{aligned}$$

$\therefore$ 1.6 kg of Batch A and 1.4 kg of Batch B were purchased.

5. 2

Let the number of rollers be x.

$\therefore$ the number of paintbrushes is $2x$ and the number of cans is $6x$.

$$\begin{aligned} 18 \times x + 16 \times 2x + 70 \times 6x &= 470 \\ 18x + 32x + 420x &= 470 \\ 470x &= 470 \\ x &= 1 \end{aligned}$$

$$\begin{aligned} \text{Also, } 2x &= 2 \times 1 \\ &= 2 \end{aligned}$$

$\therefore$ Bruce bought 2 paintbrushes.

6. 4% account: \$1800, 6% account: \$4200

Let the amount in the 4% account be x.

$\therefore$ the amount in the 6% account is $(6000 - x)$.

$$\begin{aligned} 0.04 \times x + 180 &= 0.06 \times (6000 - x) \\ 0.04x + 180 &= 360 - 0.06x \\ 0.1x &= 360 - 180 \\ 0.1x &= 180 \\ x &= 1800 \end{aligned}$$

$$\begin{aligned} \text{Also, } 6000 - x &= 6000 - 1800 \\ &= 4200 \end{aligned}$$

$\therefore$ investments were \$1800 in 4% account and \$4200 in the 6% account.

Key Skill  19 (pages 50–51)

1. \$83.30

Subs. $A = 35.7$ and $n = 3$ in $A = kn$

$$\begin{aligned} 35.7 &= 3k \\ 3k &= 35.7 \\ k &= 11.9 \end{aligned}$$

$\therefore$ the equation is $A = 11.9n$.

Subs. $n = 7$ in $A = 11.9n$

$$\begin{aligned} A &= 11.9 \times 7 \\ &= 83.3 \end{aligned}$$

$\therefore$ the cost of 7 bags is \$83.30.

2. 8.788 kg

Subs. $M = 6.912$ and $d = 2.4$ in $M = kd^3$

$$\begin{aligned} 6.912 &= k \times 2.4^3 \\ k &= \frac{6.912}{2.4^3} \\ k &= 0.5 \end{aligned}$$

$\therefore$ the equation is $M = 0.5d^3$.

Subs. $d = 2.6$ in $M = 0.5d^3$

$$\begin{aligned} &= 0.5 \times 2.6^3 \\ &= 8.788 \end{aligned}$$

$\therefore$ the mass is 8.788 kg.

3. 125 g

Subs. $E = 0.8$ and $m = 40$ in $E = km$

$$\begin{aligned} 0.8 &= 40k \\ 40k &= 0.8 \\ k &= \frac{0.8}{40} \\ k &= 0.02 \end{aligned}$$

$\therefore$ the equation is $E = 0.02m$.

Subs. $E = 2.5$ in $E = 0.02m$:

$$\begin{aligned} 2.5 &= 0.02m \\ m &= \frac{2.5}{0.02} \\ &= 125 \end{aligned}$$

$\therefore$ the mass is 125 grams.

4. 1.6 s

Subs. $T = 2$ and $L = 2500$ in $T = k\sqrt{L}$

$$2 = k \times \sqrt{2500}$$
$$50k = 2$$
$$k = 0.04$$

$\therefore$ the equation is $T = 0.04\sqrt{L}$.

Subs. $L = 1600$ in $T = 0.04\sqrt{L}$

$$= 0.04 \times \sqrt{1600}$$
$$= 1.6$$

$\therefore$ it would take 1.6 s.

5. 12 g/cm²

Subs. $P = 8$ and $d = 20$ in $P = kd$

$$8 = 20k$$
$$20k = 8$$
$$k = \frac{8}{20}$$
$$k = 0.4$$

$\therefore$ the equation is $P = 0.4d$.

$20 + 10 = 30$

Subs. $d = 30$ in $P = 0.4d$

$$= 0.4 \times 30$$
$$= 12$$

$\therefore$ the pressure is 12 g/cm².

6. Yes

Subs. $D = 3.2$ and $s = 16$ in $D = ks^2$

$$3.2 = k \times 16^2$$
$$k = \frac{3.2}{16^2}$$
$$k = 0.0125$$

$\therefore$ the equation is $D = 0.0125s^2$.

Subs. $s = 100$ in $D = 0.0125s^2$

$$= 0.0125 \times 100^2$$
$$= 125$$

It will take Sam 125 metres to stop.

$\therefore$ Yes, Sam will collide with the car.

Revision Test  5 (page 52)

1. 3x + 16

Stella's age 2 years ago $= x - 2$

'twice as old': $2(x - 2)$

'In 10 years time':

$x + 10$ and $2(x - 2) + 10$

$$\text{Sum} = x + 10 + 2(x - 2) + 10$$
$$= x + 10 + 2x - 4 + 10$$
$$= 3x + 16$$

$\therefore$ the sum is $3x + 16$.

2. 5

Let the number of 3-point shots be x.

$\therefore$ the number of 2-point shots be $(12 - x)$.

$$3x + 2(12 - x) = 29$$
$$3x + 24 - 2x = 29$$
$$x + 24 = 29$$
$$x = 5$$

$\therefore$ Shelley made five 3-point shots.

3. 43°, 51°, 86°

Let the smallest angle be $x°$.

$\therefore$ the largest angle is $2x°$, and the other angle is $(x + 8)°$.

$$x + 2x + (x + 8) = 180$$
$$4x + 8 = 180$$
$$4x = 172$$
$$x = 43$$

$2x = 2 \times 43 = 86$ and

$x + 8 = 43 + 8 = 51$

$\therefore$ angles are 43°, 51° and 86°.

4. 8:15 am

Let Abigail's time of travel be x.

$\therefore$ Noah's time of travel is $x - \frac{1}{4}$

As Distance = Speed × Time:

$$80 \times x = 100(x - \frac{1}{4})$$
$$80x = 100x - 25$$
$$20x = 25$$
$$x = \frac{25}{20}$$
$$= 1\frac{1}{4}$$

7 am plus $1\frac{1}{4}$ h = 8:15

$\therefore$ Noah caught up to Abigail at 8:15 am.

5. 4

Let the number of \$10 notes be x.

$\therefore$ the number of \$20 notes is $(x + 8)$.

$$10x + 20(x + 8) = 280$$
$$10x + 20x + 160 = 280$$
$$30x = 120$$
$$x = 4$$

$\therefore$ Ben has 4 \$10 notes in his wallet.

6. 16 cm

Let L = length in metres and m = mass in kg.

Subs. $L = 8$ and $m = 3$ in $L = km$

$$8 = 3k$$
$$3k = 8$$
$$k = \frac{8}{3}$$

$\therefore$ the equation is $L = \frac{8m}{3}$.

Subs. $m = 6$ in $L = \frac{8m}{3}$

$$L = \frac{8 \times 6}{3}$$
$$= 16$$

$\therefore$ the spring extends 16 cm.

Revision Test  6 (page 53)

1. $\frac{3x + 3y + 2z}{180}$ hours

Time = Distance ÷ Speed

First section's time: $\frac{x}{60}$ h

Break: y minutes $= \frac{y}{60}$ h

Second section's time: $\frac{z}{90}$ h

$$\text{Total time} = \frac{x}{60} + \frac{y}{60} + \frac{z}{90}$$
$$= \frac{3x + 3y + 2z}{180}$$

$\therefore$ the trip took Ava $\frac{3x + 3y + 2z}{180}$ hours.

2. \$396

Let the number of normal-time hours be x.

$\therefore$ the number of double-time hours is $(x + 2)$.

$$x + (x + 2) = 14$$
$$2x + 2 = 14$$
$$2x = 12$$
$$x = 6$$

Lucas worked 6 hours normal time and 8 hours double time.

$$\text{Total pay} = 18 \times 6 + 18 \times 2 \times 8$$
$$= 396$$

$\therefore$ Lucas earned \$396.

3. \$14.70

Let the width be x.

$\therefore$ the length is $(2x - 8)$.

$$\text{Perimeter} = 2(2x - 8 + x)$$
$$= 2(3x - 8)$$

Also, $67 \div 50 = 1.34$.

1.34 m = 134 cm.

$$\therefore \quad 2(3x - 8) = 134$$
$$6x - 16 = 134$$
$$6x = 150$$
$$x = 25$$

Now, $2x - 8 = 2(25) - 8$

$$= 42$$

Dimensions are 42 cm by 25 cm, or 0.42 m by 0.25 m.

$$\text{Area} = 0.42 \times 0.25$$
$$= 0.105$$
$$\text{Cost of glass} = 0.105 \times 140$$
$$= 14.70$$

$\therefore$ the cost is \$14.70.

4. **2 km**

$40 \text{ min} = \frac{2}{3} \text{ h}$

Let the distance from apartment to shops be x.

As Time = Distance ÷ Speed:

$\frac{x}{6} + \frac{x}{12} + \frac{x}{12} = \frac{2}{3}$

$12 \times \frac{x}{6} + 12 \times \frac{x}{12} + 12 \times \frac{x}{12}$

$= 12 \times \frac{2}{3}$

$2x + x + x = 8$

$4x = 8$

$x = 2$

∴ Candice lives 2 km from the shops.

5. **\$14 000, \$6000**

Let the money in the first account be \$$x$.

∴ the money in second account is \$$(20\,000 - x)$.Using $I = Prn$:

$4425 = x \times 0.045 \times 5$
$+ (20\,000 - x) \times 0.0425 \times 5$

$4425 = 0.225x + 0.2125(20\,000 - x)$

$4425 = 0.225x + 4250 - 0.2125x$

$0.0125x = 175$

$x = \frac{175}{0.0125}$

$= \frac{175}{0.0125}$

$= 14\,000$

$20\,000 - x = 20\,000 - 14\,000$

$= 6000$

∴ the two amounts were \$14 000 and \$6000.

6. **2 h 15 min**

Let N = number of eggs and h = hours slept.

Subs. $N = 4$ and $h = 9$ in $N = k\sqrt{h}$:

$4 = k\sqrt{9}$

$3k = 4$

$k = \frac{4}{3}$

∴ the equation is $N = \frac{4\sqrt{h}}{3}$.

Subs. $N = 2$ in $N = \frac{4\sqrt{h}}{3}$

$2 = \frac{4\sqrt{h}}{3}$

$4\sqrt{h} = 6$

$\sqrt{h} = \frac{3}{2}$

$h = \frac{9}{4}$

$= 2 \text{ h } 15 \text{ min}$

∴ the goose sleeps for 2 h 15 min.

Key Skill  20 (pages 54–55)

1. **Less than 8 years old**

Let Emma's age be x.

∴ Sophia's age is $x + 3$.

$x + x + 3 < 19$

$2x + 3 < 19$

$2x < 19 - 3$

$2x < 16$

$x < 8$

∴ Emma must be less than 8 years old.

2. **2**

Let the number of shirts be x.

$200 - 40x \geq 90$

$-40x \geq 90 - 200$

$-40x \geq 90 - 200$

$-40x \geq -110$

$\frac{-40x}{-40} \leq \frac{-110}{-40}$

$x \leq 2.75$

∴ Angus can buy 2 shirts.

3. **32**

Let Jacob's age be x.

∴ William's age is $x - 5$, Ethan's age is $x + 8$.

$x + x - 5 + x + 8 \leq 99$

$3x + 3 \leq 99$

$3x \leq 99 - 3$

$3x \leq 96$

$x \leq 32$

∴ Jacob's maximum age is 32.

4. **31**

Let the number of shower screens sold be x.

$120x > 2400 + 40x$

$120x - 40x > 2400$

$80x > 2400$

$x > 30$

∴ the factory has to sell at least 31 shower screens.

5. **6**

Let the number of weeks be x.

$5400 - 640x \geq 1300$

$-640x \geq 1300 - 5400$

$-640x \geq -4100$

$\frac{-640x}{-640} \leq \frac{-4100}{-640}$

$x \leq 6.40625$

∴ Alexis's money will last 6 weeks.

6. **15**

Let the number of trailers be x.

Selling price − Cost price = Profit.

$3200x - (6000 + 2100x) \geq 10\,000$

$3200x - 6000 - 2100x \geq 10\,000$

$1100x \geq 10\,000 + 6000$

$1100x \geq 16\,000$

$\frac{1100x}{1100} \geq \frac{16\,000}{1100}$

$x \geq 14.545\ldots$

∴ Sean needs to sell 15 trailers.

Key Skill  21 (pages 56–57)

1. **20**

Subs. $V = 12$, $E = 30$ in

$V + F = E + 2$

$12 + F = 30 + 2$

$12 + F = 32$

$F = 32 - 12$

$= 20$

∴ an icosohedron has 20 faces.

2. **12**

Subs. $V = 60$, $E = 90$ in

$V + F = E + 2$

$60 + F = 90 + 2$

$60 + F = 92$

$F = 92 - 60$

$= 32$

As $32 - 20 = 12$, it must have 12 pentagonal faces.

∴ a bucky-ball has 12 pentagonal faces.

3. **125 m**

Subs. $g = 10$, $t = 5$ in

$h = \frac{1}{2}gt^2$

$h = \frac{1}{2} \times 10 \times 5^2$

$= 125$

∴ object will fall 125 metres.

4. **14.1 s**

Subs. $h = 1000$, $g = 10$ in

$h = \frac{1}{2}gt^2$

$1000 = \frac{1}{2} \times 10 \times t^2$

$5t^2 = 1000$

$t^2 = 200$

$t = \sqrt{200} \quad (t > 0)$

$= 14.1$ (3 sig. figs)

∴ it will take 14.1 s.

5. **4.9 s**

Subs. $L = 6$, $g = 10$ in

$P = 2\pi\sqrt{\frac{L}{g}}$

$= 2\pi \times \sqrt{\frac{6}{10}}$

$= 4.866\,934\,411\ldots$

$= 4.9$ (1 dec. pl.)

∴ it will take 4.9 s.

6. **16 m**
Subs. $P = 8, g = 10$ in

$$P = 2\pi\sqrt{\frac{L}{g}}$$
$$8 = 2\pi \times \sqrt{\frac{L}{10}}$$
$$\sqrt{\frac{L}{10}} = \frac{4}{\pi}$$
$$\frac{L}{10} = \frac{16}{\pi^2}$$
$$L = \frac{160}{\pi^2}$$
$$= 16.211\,389\,38\ldots$$
$$= 16 \text{ (nearest whole)}$$

∴ the pendulum is 16 metres long.

Key Skill 22 (pages 58–59)

1. **\$108**
$$C = 6n$$
$$= 6 \times 18$$
$$= 108$$
∴ the cost is \$108.

2. **\$110**
$$C = 30 + 20t$$
$$= 30 + 20 \times 4$$
$$= 110$$
∴ the cost is \$110.

3. **2**
$$c = 50 - 3n$$
$$= 50 - 3 \times 16$$
$$= 2$$
∴ Inger will have 2 chocolates remaining.

4. **\$81.25**
$$C = 6.25 + 3d$$
$$= 6.25 + 3 \times 25$$
$$= 81.25$$
∴ it will cost \$81.25.

5. **57.5 km**
$$d = 620 - 90n$$
$$= 620 - 90 \times 6.25$$
$$= 57.5$$
∴ Anatoly still has 57.5 km to drive.

6. **1.5 km**

n	1	2	5	8
D	300	350	500	650

$$D = 250 + 50n$$
$$= 250 + 50 \times 25$$
$$= 1500$$
∴ Sarah swam 1500 m, or 1.5 km.

Key Skill 23 (pages 60–61)

1. **Joachin 14, Leila 8**
Let Joachin's age be x and Leila's age be y.
$$x + y = 22 \ldots ①$$
$$x - y = 6 \ldots ②$$
① + ②: $2x = 28$
$$x = 14$$
Subs. in ① :$14 + y = 22$
$$y = 8$$
∴Joachin is 14 and Leila is 8.

2. **4000 cm²**
Let the width be x cm and the length be y cm.
$$2(x + y) = 260$$
i.e. $x + y = 130 \ldots ①$
Also, $y = 2x - 20 \ldots ②$
Subs ② in ① : $x + 2x - 20 = 130$
$$3x - 20 = 130$$
$$3x = 150$$
$$x = 50$$
Subs. in ① : $y = 2(50) - 20$
$$y = 80$$
∴ the dimensions are 80 cm and 50 cm.
∴ the area is 4000 cm².

3. **24 sheep**
Let the number of chickens be x and the number of sheep be y.
$$x + y = 40 \ldots ①$$
$$2x + 4y = 128$$
$$x + 2y = 64 \ldots ②$$
② − ①: $y = 24$
∴ there are 24 sheep.

4. **mother: 34, daughter: 8**
Let the daughter's age be x, the mother's age be y.
$$2x + y = 50 \ldots ①$$
$$5x - y = 6 \ldots ②$$
① + ②: $7x = 56$
$$x = 8$$
Subs. in ①: $2(8) + y = 50$
$$16 + y = 50$$
$$y = 34$$
∴ the mother is 34 and the daughter is 8.

5. **3 buses, 7 cars**
Let the number of buses be x, the number of cars be y.
$$x + y = 10 \ldots ①$$
$$46x + 4y = 166 \ldots ②$$
From ①: $y = 10 - x \ldots ③$
Subs ③ in ②:
$$46x + 4(10 - x) = 166$$
$$46x + 40 - 4x = 166$$
$$42x + 40 = 166$$
$$42x = 126$$
$$x = 3$$
Subs. in ①: $3 + y = 10$
$$y = 7$$
∴ there were 3 buses and 7 cars.

6. **Model A: 5, Model B: 3**
Let the number of Model A be a, the number of Model B be b.
$$a + b = 8 \ldots ①$$
$$1200a + 1000b = 9000 \ldots ②$$
From ①: $b = 8 - a \ldots ③$
Subs ③ in ②:
$$1200a + 1000(8 - a) = 9000$$
$$1200a + 8000 - 1000a = 9000$$
$$200a + 8000 = 9000$$
$$200a = 1000$$
$$a = 5$$
Subs. in ①: $5 + b = 8$
$$b = 3$$
∴ there were 5 Model A printers and 3 Model B printers.

Key Skill 24 (pages 62–63)

1. **milkshake: \$5.60, juice: \$3.80**
Let the cost of a milkshake be x and the cost of a juice be y.
$$3x + 2y = 24.4 \ldots ①$$
$$2x + 3y = 22.6 \ldots ②$$
3 × ①: $9x + 6y = 73.2 \ldots ③$
2 × ②: $4x + 6y = 45.2 \ldots ④$
③ − ④: $5x = 28$
$$x = 5.6$$
Subs in ①: $3(5.6) + 2y = 24.4$
$$16.8 + 2y = 24.4$$
$$2y = 7.6$$
$$y = 3.8$$
∴ the milkshake costs \$5.60 and the juice costs \$3.80.

2. **jogs: 12 km/h, walks: 6 km/h**
Let jogging speed be x and walking speed be y.
As Distance = Speed × Time:
$$\frac{1}{2}x + \frac{1}{3}y = 8$$
$$3x + 2y = 48 \ldots ①$$
$$\frac{1}{3}x + \frac{1}{2}y = 7$$
$$2x + 3y = 42 \ldots ②$$
3 × ①: $9x + 6y = 144 \ldots ③$
2 × ②: $4x + 6y = 84 \ldots ④$
③ − ④: $5x = 60$
$$x = 12$$
Subs. in ①: $3(12) + 2y = 48$
$$36 + 2y = 48$$
$$2y = 12$$
$$y = 6$$
∴Jodie jogs at 12 km/h and walks at 6 km/h.

3. 12

Let the number of 5-cent coins be x.
$\therefore$ the number of 10-cent coins be $2x$.
Let the number of 20-cent coins be y.

$$x + 2x + y = 60$$
$$3x + y = 60 \ldots ①$$
$$5 \times x + 10 \times 2x + 20 \times y = 640$$
$$25x + 20y = 640 \ldots ②$$
$$25 \times ①: 75x + 25y = 1500 \ldots ③$$
$$3 \times ②: 75x + 60y = 1920 \ldots ④$$
$$④ - ③: 35y = 420$$
$$y = 12$$

$\therefore$ there are twelve 20-cent coins.

4. 12 km/h, 8 km/h

Let the speed of first runner be x, and the speed of the second runner be y.
Let $x > y$.
As Distance = Speed × Time:

$$2x - 2y = 8$$
$$x - y = 4 \ldots ①$$
$$x + y = 20 \ldots ②$$
$$② - ①: 2y = 16$$
$$y = 8$$

Subs in ①: $x - 8 = 4$
$$x = 12$$

$\therefore$ the speeds are 12 km/h and 8 km/h.

5. 200

Let the number of pensioners be x.
$\therefore$ the number of children is $3x$.
Let the number of adults be y.

$$x + 3x + y = 360$$
$$4x + y = 360 \ldots ①$$
$$14 \times x + 10 \times 3x + 20 \times y = 5760$$
$$44x + 20y = 5760 \ldots ②$$
$$11 \times ①: 44x + 11y = 3960 \ldots ③$$
$$② - ③: 9y = 1800$$
$$y = 200$$

$\therefore$ 200 adult tickets were sold.

6. 2.5 km

Let the distance ridden be x and the distance walked be y.

$$x + y = 20 \ldots ①$$

As Time = Distance ÷ Speed

$$\frac{x}{30} + \frac{y}{6} = 1$$
$$x + 5y = 30 \ldots ②$$
$$② - ①: 4y = 10$$
$$y = 2.5$$

$\therefore$ Enrico walked 2.5 km.

Key Skill  25 (pages 64–65)

1. 8 and 4

Let the age of Gemma be x and the age of Phoebe be y.

$x + y = 12$
$\therefore y = 12 - x$

x	5	6	7
y	7	6	5

$x - y = 4$
$\therefore y = x - 4$

x	5	6	7
y	1	2	3

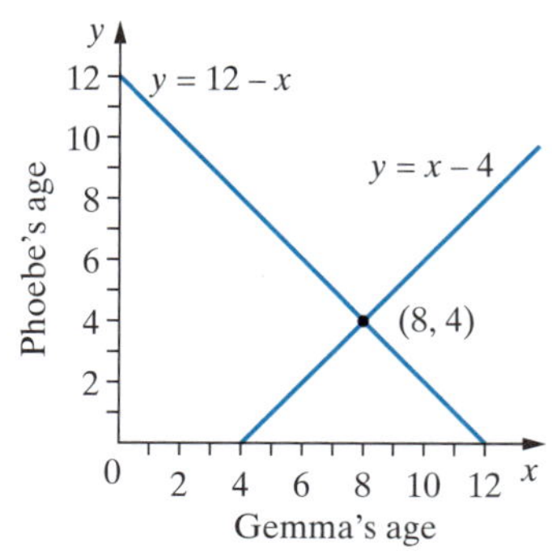

$\therefore$ Gemma is 8 and Phoebe is 4.

2. 3 hours

Let C be the charge (in \$) and n be the number of hours.

Barry: $C = 60 + 40n$

n	0	1	2
C	60	100	140

Garry: $C = 60n$

n	0	1	2
C	0	60	120

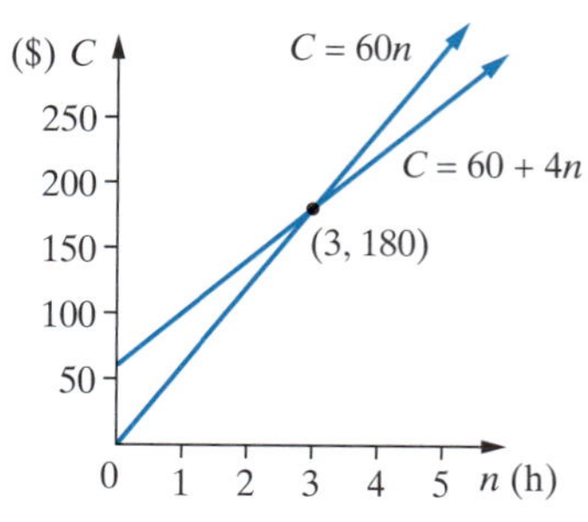

$\therefore$ The electricians are paid the same amount for a 3-hour job.

3. 5 weeks

Let S be the savings (in \$) and n be the number of weeks.

Simone: $S = 600 + 100n$

n	0	1	2
S	600	700	800

Kate: $S = 500 + 120n$

n	0	1	2
S	500	620	740

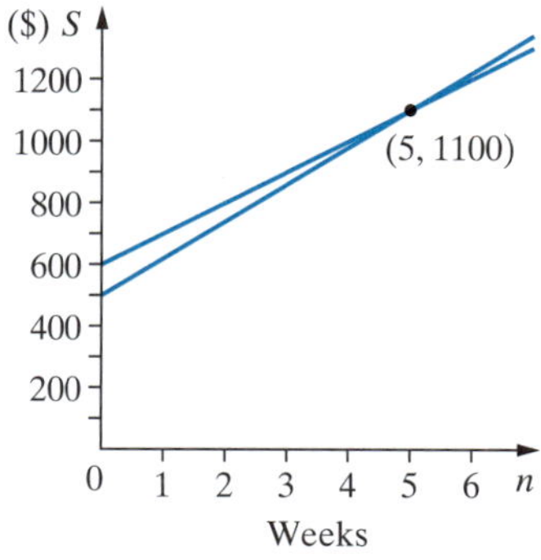

$\therefore$ After 5 weeks the girls have saved the same amount.

4. \$8 and \$4

Let a be the cost (in \$) of an adult ticket and c be the cost (in \$) of a child ticket.

$$2a + c = 20$$
$$c = 20 - 2a$$

a	0	2	4
c	20	16	12

Also, $a + 4c = 24$

$$4c = 24 - a$$
$$c = \frac{24 - a}{4}$$

a	0	2	4
c	6	5.5	5

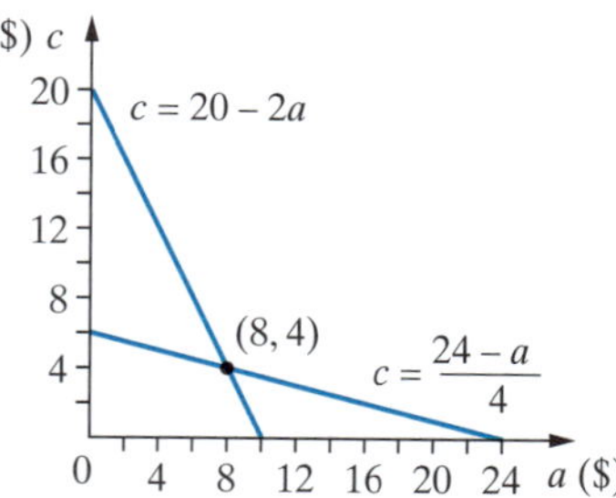

$\therefore$ an adult ticket costs \$8 and a child ticket costs \$4.

5. 2 min 30 s

Let A be the amount of water (in mL) and n be the number of minutes.

Bucket A: $A = 6000 - 400n$

n	0	1	2
A	6000	5600	5200

Bucket B: $A = 4000 + 400n$

n	0	1	2
A	4000	4400	4800

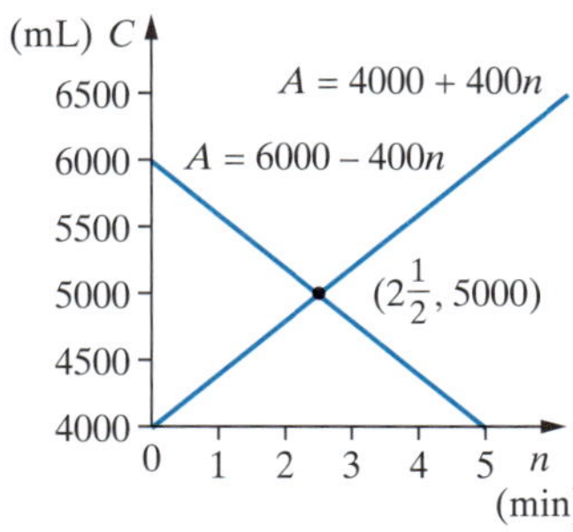

$\therefore$ the buckets have the same amount after 2 min 30 s.

6. 12:30 pm

Let D be the distance travelled (in km) and t be the time (in hours).

Ethan: $D = 12t$

t	1	2	3
D	12	24	36

Jake: $D = 16(t - \frac{3}{4})$

$D = 16t - 12$

t	1	2	3
D	4	20	36

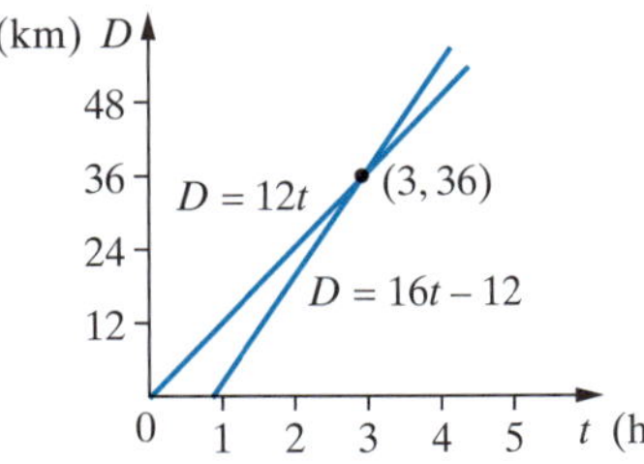

$\therefore$ Ethan passes Jake after 3 h.

9:30 am + 3 h = 12:30 pm

$\therefore$ Ethan passes Jake at 12:30 pm.

Key Skill  26 (pages 66–67)

1. −13 and −11 or 11 and 13

Let the two numbers be x and $x + 2$.

$$x(x + 2) = 143$$
$$x^2 + 2x = 143$$
$$x^2 + 2x - 143 = 0$$
$$(x + 13)(x - 11) = 0$$
$$x = -13 \text{ or } x = 11$$

As $x = -13$, then
$x + 2 = -13 + 2 = -11$

As $x = 11$, then
$x + 2 = 11 + 2 = 13$

$\therefore$ the numbers are -13 and -11 or 11 and 13.

2. 38 cm

Let the width be x.

Let the length be $x + 5$.

$$x(x + 5) = 84$$
$$x^2 + 5x - 84 = 0$$
$$(x + 12)(x - 7) = 0$$
$$x = -12, 7$$
$$x = 7 \quad (\text{As } x > 0)$$

As $x = 7$, $x + 5 = 12$.

The dimensions are 12 cm and 7 cm.

$$\text{Perimeter} = 2(12 + 7)$$
$$= 38$$

$\therefore$ the perimeter is 38 cm.

3. 6 and 8

Let one number be x.

Let the other number be $14 - x$.

$$x^2 + (14 - x)^2 = 100$$
$$x^2 + 196 - 28x + x^2 = 100$$
$$2x^2 - 28x + 96 = 0$$
$$x^2 - 14x + 48 = 0$$
$$(x - 6)(x - 8) = 0$$
$$x = 6, 8$$

As $x = 6$, $14 - x = 8$.

$\therefore$ the two numbers are 6 and 8.

4. \$568.40

Let the length be x.

Let the width be $x - 2$.

$$x(x - 2) = 2400$$
$$x^2 - 2x - 2400 = 0$$
$$(x - 50)(x + 48) = 0$$
$$x = 50, -48$$
$$x = 50 \quad (\text{As } x > 0)$$

As $x = 50$, $x - 2 = 48$.

The dimensions are 50 m and 48 m.

$$\text{Perimeter} = 2(50 + 48)$$
$$= 196$$
$$\text{Cost} = 2.9 \times 196$$
$$= 568.4$$

$\therefore$ the fence cost is \$568.40.

5. 2 m

8 + 2x
8
14
x
14 + 2x

Let the width of the path be x.

The dimensions of the total area are $(14 + 2x)$ cm and $(8 + 2x)$ cm.

$$(2x + 14)(2x + 8) = 216$$
$$4x^2 + 44x + 112 = 216$$
$$4x^2 + 44x - 104 = 0$$
$$x^2 + 11x - 26 = 0$$
$$(x + 13)(x - 2) = 0$$
$$x = -13, 2$$
$$= 2 \quad (x > 0)$$

$\therefore$ the width of the grass path is 2 m.

6. 5

Let the number be x.

$$\sqrt{x - 1} = 7 - x$$
$$x - 1 = (7 - x)^2$$
$$x - 1 = 49 - 14x + x^2$$
$$x^2 - 15x + 50 = 0$$
$$(x - 5)(x - 10) = 0$$
$$x = 5, 10$$

The two possible answers are 5 and 10.

However, subs. $x = 10$:
$\sqrt{10 - 1} \neq 7 - 10$

$\therefore$ the number is 5.

Key Skill  27 (pages 68–69)

1. base: 24 cm

Let perpendicular height be x.

$\therefore$ the length of the base is $2x + 4$.

$$\frac{1}{2}x(2x + 4) = 120$$
$$x^2 + 2x = 120$$
$$x^2 + 2x - 120 = 0$$
$$(x + 12)(x - 10) = 0$$
$$x = -12 \text{ or } x = 10$$
$$x = 10 \quad (\text{As } x > 0)$$

As $x = 10$, then
$2x + 4 = 2(10) + 4 = 24$.

$\therefore$ the base is 24 cm.

2. **24 cm²**

Let the sides be x, $(x-2)$, $(x+2)$.

By Pythagoras:

$$x^2 + (x-2)^2 = (x+2)^2$$
$$x^2 + x^2 - 4x + 4 = x^2 + 4x + 4$$
$$2x^2 - 4x + 4 = x^2 + 4x + 4$$
$$x^2 - 8x = 0$$
$$x(x-8) = 0$$
$$x = 0, 8$$
$$x = 8 \quad (\text{as } x \neq 0)$$

As $x = 8$, $x - 2 = 6$ and $x + 2 = 10$.

∴ the dimensions are 6 cm, 8 cm, 10 cm.

$$\text{Area} = \frac{1}{2} \times 6 \times 8$$
$$= 24$$

∴ the area is 24 cm².

3. **$7\frac{1}{3}$ cm**

Let length of rectangle be $2x$.

∴ the width of rectangle is $(x-1)$.

$$2x(x-1) = 24$$
$$2x^2 - 2x - 24 = 0$$
$$x^2 - x - 12 = 0$$
$$(x-4)(x+3) = 0$$
$$x = 4, -3$$
$$x = 4 \quad (\text{as } x > 0)$$

As $x = 4$, then $2x = 2(4) = 8$ and $x - 1 = 4 - 1 = 3$

∴ the dimensions are 8 cm and 3 cm.

$$\text{Perimeter} = 2(8+3)$$
$$= 22$$
$$\text{Side length of triangle} = \frac{22}{3}$$
$$= 7\frac{1}{3}$$

∴ each side is $7\frac{1}{3}$ cm.

4. **48 cm**

Let sides of the square be x.

Area of square $= x^2$.

∴ sides of rectangle are $(x+6)$ and $(x-4)$.

Area of rectangle $= (x+6)(x-4)$

$$(x+6)(x-4) = x^2$$
$$x^2 + 2x - 24 = x^2$$
$$2x = 24$$
$$x = 12$$

As $x = 12$, then $4x = 4(12) = 48$.

∴ the perimeter was 48 cm.

5. **28 cm²**

Let side of the smaller square be x.

Area of smaller square $= x^2$.

∴ side of larger square is $(x+2)$.

Area of larger square $= (x+2)^2$.

$$x^2 + (x+2)^2 = 100$$
$$x^2 + x^2 + 4x + 4 - 100 = 0$$
$$2x^2 + 4x - 96 = 0$$
$$x^2 + 2x - 48 = 0$$
$$(x+8)(x-6) = 0$$
$$x = -8, 6$$
$$x = 6 \quad (\text{as } x > 0)$$

As $x = 6$, then $x + 2 = 6 + 2 = 8$

Difference in area $= 8^2 - 6^2$

$= 28$

∴ the difference is 28 cm².

6. **180 cm²**

Let length of rectangle be x.

∴ width of rectangle is $(x-8)$.

After removal of squares the dimensions of the box are $(x-4)$, $(x-12)$ and 2.

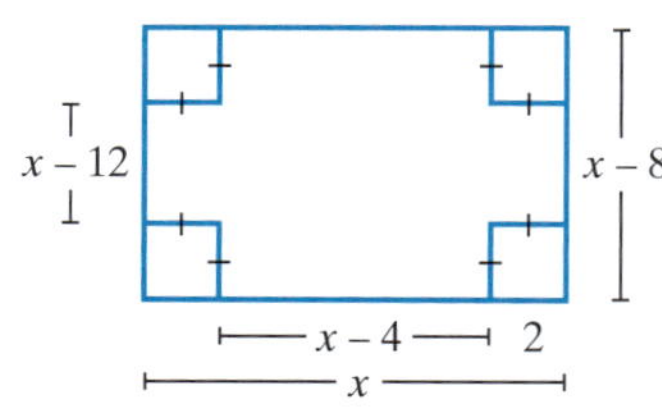

$$2(x-4)(x-12) = 168$$
$$(x-4)(x-12) = 84$$
$$x^2 - 16x + 48 - 84 = 0$$
$$x^2 - 16x - 36 = 0$$
$$(x-18)(x+2) = 0$$
$$x = 18, -2$$
$$x = 18 \quad (\text{as } x > 0)$$

As $x = 18$, then

$x - 8 = 18 - 8 = 10$

Area $= 18 \times 10$

$= 180$

∴ the area was 180 cm².

Revision Test  7 (page 70)

1. **8**

Let the number of rides be x.

$$3x + 7 \leq 32$$
$$3x \leq 32 - 7$$
$$3x \leq 25$$
$$x \leq \frac{25}{3}$$
$$x \leq 8\frac{1}{3}$$

∴ Donald can go on 8 rides.

2. **$2\frac{2}{3}$ cm**

Subs. $S = \frac{64\pi}{3}$ in $S = 3\pi r^2$

$$\frac{64\pi}{3} = 3\pi r^2$$
$$\pi r^2 = \frac{64\pi}{9}$$
$$r^2 = \frac{64}{9}$$
$$r = \frac{8}{3} \quad (r > 0)$$
$$= 2\frac{2}{3}$$

∴ the radius is $2\frac{2}{3}$ cm.

3. **80 mL**

$A = 400 - 20n$

Subs. $n = 16$ in $A = 400 - 20n$:

$$A = 400 - 20(16)$$
$$= 80$$

∴ there is 80 mL in the bottle.

4. **3**

Let the number of 3D films be x and the number of 2D films be y.

$$x + y = 10 \ldots ①$$
$$18x + 15y = 159 \ldots ②$$

15 × ①: $15x + 15y = 150 \ldots ③$

② − ③: $3x = 9$

$x = 3$

∴ Pablo saw 3 3D films.

5. **4**

Let D be the distance run in metres, and n be the number of weeks to run.

Avery: $D = 2000 + 500n$

n	0	1	2
D	2000	2500	3000

Summer: $D = 1200 + 700n$

n	0	1	2
D	1200	1900	2600

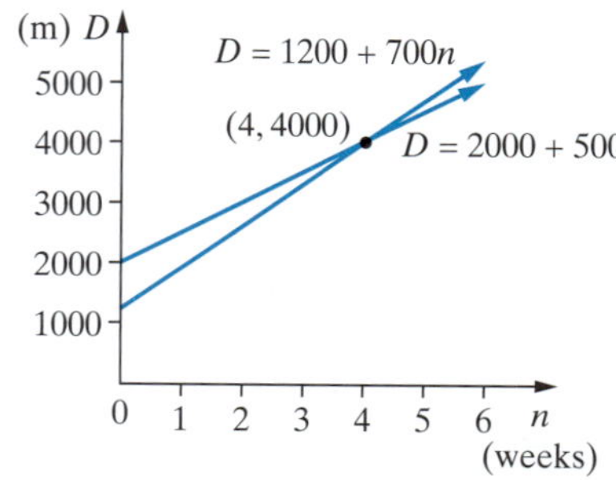

∴ in 4 weeks they run the same distance.

6. **20 m by 18 m**

Let the width be x metres.

∴ the length is $(x+2)$ metres.

$$x(x+2) = 360$$
$$x^2 + 2x - 360 = 0$$
$$(x+20)(x-18) = 0$$
$$x = -20, 18$$
$$x = 18 \quad (x > 0)$$

$x + 2 = 18 + 2 = 20$

∴ the dimensions are 20 m by 18 m.

Revision Test  (page 71)

1. 19

Let the number of lawnmowers sold be x.

Selling price − Cost price = Profit

$$210x - (420 + 80x) \geq 2000$$
$$210x - 420 - 80x \geq 2000$$
$$130x \geq 2420$$
$$x \geq \frac{2420}{130}$$
$$x \geq 18\frac{8}{13}$$

∴ Sean needs to sell at least 19 lawnmowers.

2. dress \$180, jeans \$80

Let the price paid for the dress be x and the price paid for the jeans be y.

$x + y = 260$ … ①

$0.9x + 0.85y = 230$ … ②

9 × ①: $9x + 9y = 2340$ … ③

10 × ②: $9x + 8.5y = 2300$ … ④

③ − ④: $0.5y = 40$

$y = 80$

Subs. in ①: $x + 80 = 260$

$x = 180$

∴ Fiona paid \$180 for the dress and \$80 for the jeans.

3. 38.025

Subs. $t = 1.5$, $u = 18$, $a = 9.8$ in

$$t = \frac{\sqrt{u^2 + 2as} - u}{a}$$
$$1.5 = \frac{\sqrt{18^2 + 2(9.8)s} - 18}{9.8}$$
$$14.7 = \sqrt{324 + 19.6s} - 18$$
$$\sqrt{324 + 19.6s} = 32.7$$
$$324 + 19.6s = 1069.29$$
$$19.6s = 745.29$$
$$s = \frac{745.29}{19.6}$$
$$= 38.025$$

4. 40 cm

Let the diagonals be x cm and $(x + 4)$ cm.

$$\frac{1}{2}x(x + 4) = 96$$
$$x^2 + 4x = 192$$
$$x^2 + 4x - 192 = 0$$
$$(x + 16)(x - 12) = 0$$
$$x = -16, 12$$
$$x = 12 \quad (x > 0)$$

$x + 4 = 12 + 4 = 16$

∴ the diagonals are 12 cm and 16 cm.

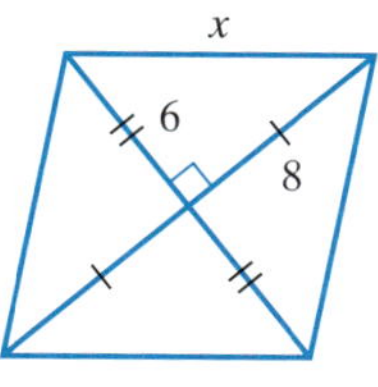

$$x^2 = 6^2 + 8^2$$
$$= 100$$
$$x = 10$$
$$\text{Perimeter} = 4 \times 10$$
$$= 40$$

∴ the perimeter is 40 cm.

5. 4 or −1

Let the denominator be x and the numerator be $(x + 4)$.

$$\frac{x + 4}{x} = x - 2$$
$$x + 4 = x(x - 2)$$
$$x^2 - 2x = x + 4$$
$$x^2 - 3x - 4 = 0$$
$$(x - 4)(x + 1) = 0$$
$$x = 4, -1$$

∴ the denominators could be 4 or −1.

6. 7:32 am

Let A be the amount of sand (in cm^3) and n be the number of minutes.

Upper container: $A = 960 - 15n$

n	0	5	10	15
A	960	885	810	735

Lower container: $A = 600 + 15n$

n	0	5	10	15
A	600	675	750	825

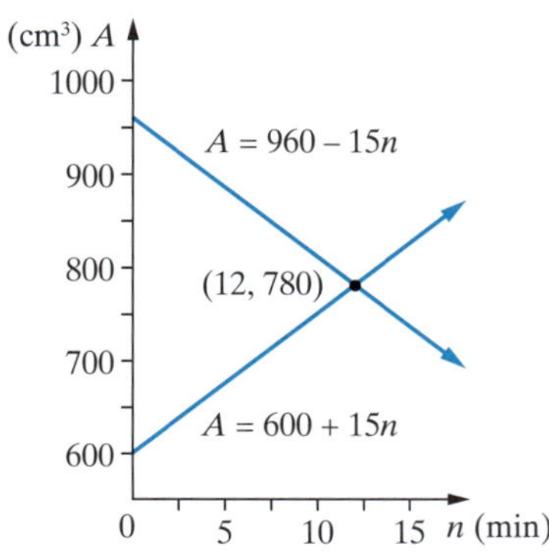

The graphs intersect when $n = 12$.

7:20 plus 12 min is 7:32.

∴ the containers have the same amount at 7:32 am.

Key Skill  (pages 72–73)

1. 18π cm²

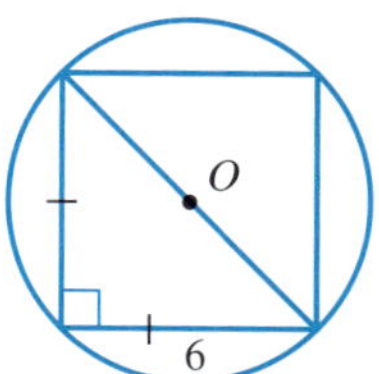

Let diameter of circle be x.

$$x^2 = 6^2 + 6^2$$
$$= 72$$
$$x = \sqrt{72}$$

∴ radius is $\frac{\sqrt{72}}{2}$ cm.

$$\text{Area} = \pi r^2$$
$$= \pi \times \left(\frac{\sqrt{72}}{2}\right)^2$$
$$= \pi \times 18$$
$$= 18\pi$$

∴ the area is 18π cm^2.

2. 111 cm²

Let height of triangle be x.

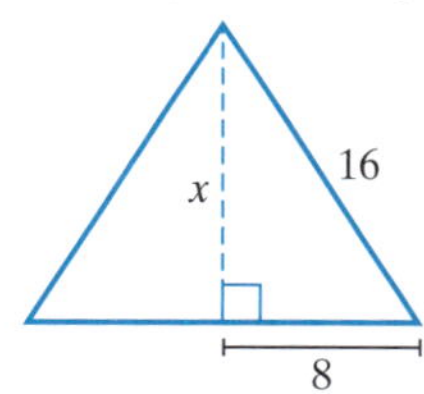

$$x^2 = 16^2 - 8^2$$
$$= 192$$
$$x = \sqrt{192}$$
$$\text{Area} = \frac{1}{2} \times 16 \times \sqrt{192}$$
$$= 110.851\,2517\ldots$$
$$= 111 \text{ (to 3 sig. figs.)}$$

∴ the area is 111 cm^2.

3. 14 min

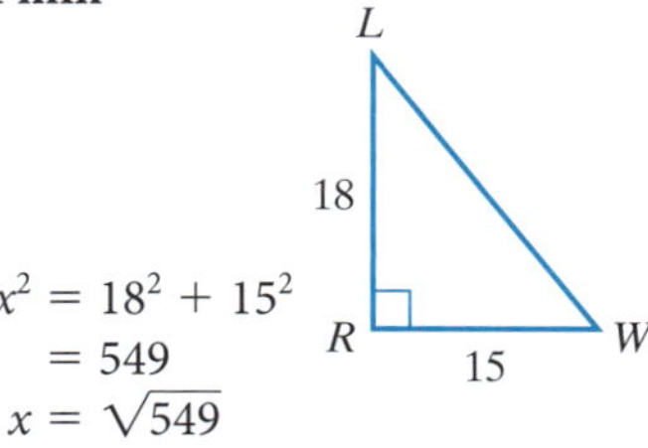

$$x^2 = 18^2 + 15^2$$
$$= 549$$
$$x = \sqrt{549}$$

Difference in distance

$$= 18 + 15 - \sqrt{549}$$
$$= 9.569\,250\,972\ldots$$
$$= 9.569 \text{ (3 dec. pl.)}$$

$$\text{Time} = \text{Distance} \div \text{Speed}$$
$$= 9.569 \div 40$$
$$= 0.239\,231\,274\ldots \text{ h}$$
$$= 14 \text{ min } 21 \text{ s}$$

∴ Ryan would save 14 min.

4. \$2116

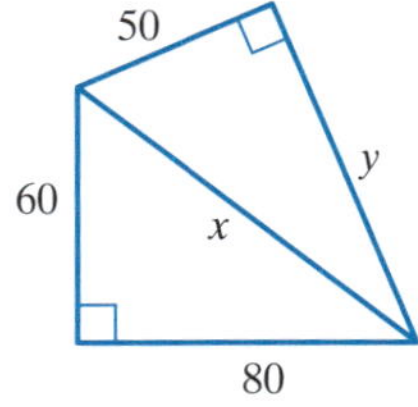

$$x^2 = 60^2 + 80^2$$
$$= 10\,000$$
$$y^2 = 10\,000 - 50^2$$
$$= 7500$$
$$y = \sqrt{7500}$$

Perimeter

$$= 80 + 60 + 50 + \sqrt{7500}$$
$$= 276.602\,5404\ldots$$
$$= 276.603 \text{ (3 dec. pl.)}$$

Cost $= 276.603 \times 7.65$

$= 2116$ (nearest whole)

$\therefore$ the cost is \$2116.

5. 864 m²

Let half of unknown diagonal be x.

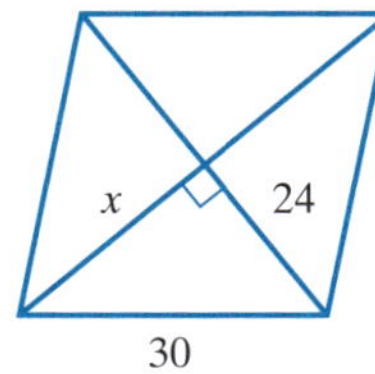

$$x^2 = 30^2 - 24^2$$
$$= 324$$
$$x = 18$$

Diagonal $= 2 \times 18$

$= 36$

Area $= \frac{1}{2}$ product of diagonals

$$= \frac{1}{2} \times 48 \times 36$$
$$= 864$$

$\therefore$ the area is 864 m².

6. 46 m²

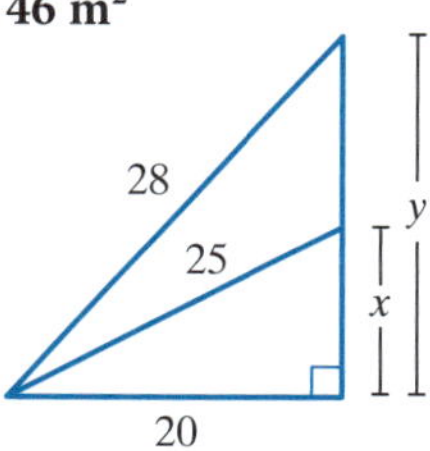

$$y^2 = 28^2 - 20^2$$
$$= 384$$
$$y = \sqrt{384}$$
$$x^2 = 25^2 - 20^2$$
$$= 225$$
$$x = 15$$

Sign width $= \sqrt{384} - 15$

Sign area $= 10(\sqrt{384} - 15)$

$= 45.959\,179\,42\ldots$

$= 46$ (nearest whole)

$\therefore$ the area is 46 m².

Key Skill  (pages 74–75)

1. 100 km

Distance = Speed × Time

Boat X: Distance $= 15 \times 4$

$= 60$

Boat Y: Distance $= 20 \times 4$

$= 80$

Let $XY = x$

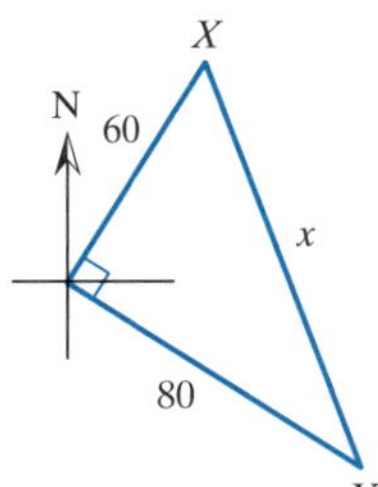

$$x^2 = 60^2 + 80^2$$
$$= 10\,000$$
$$x = 100$$

$\therefore$ the boats are 100 km apart.

2. 60 cm²

Let the shortest side be x.

$\therefore$ the longest side is $(2x + 1)$ and the other side is $(2x - 1)$.

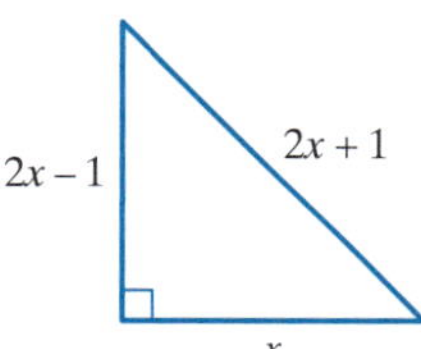

$$x^2 + (2x - 1)^2 = (2x + 1)^2$$
$$x^2 + 4x^2 - 4x + 1 = 4x^2 + 4x + 1$$
$$x^2 - 8x = 0$$
$$x(x - 8) = 0$$
$$x = 0, 8$$
$$x = 8 \quad (\text{as } x > 0)$$

Dimensions are 8 cm, 15 cm and 17 cm.

$$\text{Area} = \frac{1}{2} \times 8 \times 15$$
$$= 60$$

$\therefore$ the area is 60 cm².

3. $(2\pi - 4)$ cm²

Let radius be x.

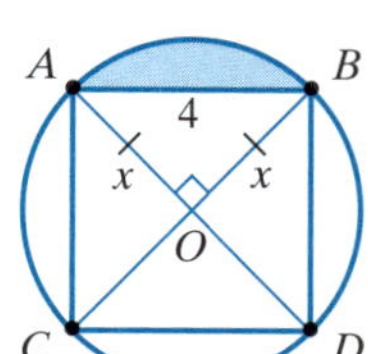

$$x^2 + x^2 = 4^2$$
$$2x^2 = 16$$
$$x^2 = 8$$

Area of quadrant $= \frac{1}{4}\pi r^2$

$$= \frac{1}{4}\pi \times x^2$$
$$= \frac{1}{4}\pi \times 8$$
$$= 2\pi$$

Area of triangle $= \frac{1}{2} \times x \times x$

$$= \frac{1}{2}x^2$$
$$= \frac{1}{2} \times 8$$
$$= 4$$

Shaded area = Area of quadrant − area of triangle

$$= 2\pi - 4$$

$\therefore$ the area is $(2\pi - 4)$ cm².

4. 0.8 m

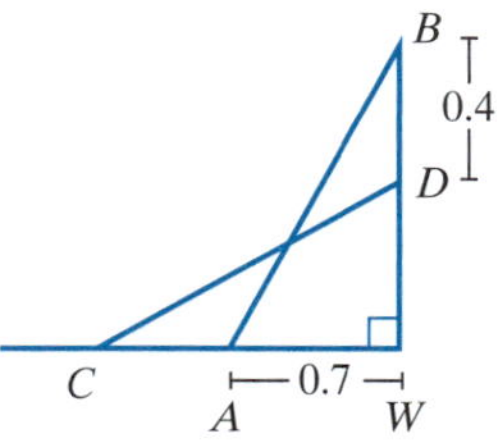

Ladder $= AB = CD = 2.5$

Let $BW = x$.

$$x^2 = 2.5^2 - 0.7^2$$
$$= 5.76$$
$$x = 2.4$$

$\therefore DW = 2.4 - 0.4 = 2$

Let $CW = y$

$$y^2 = 2.5^2 - 2^2$$
$$= 2.25$$
$$y = 1.5$$

$\therefore CA = 1.5 - 0.7 = 0.8$

$\therefore$ the ladder moved out 0.8 m.

5. 50 km

Time for Yacht A $= 30 \div 12$

$= 2.5$

$\therefore$ yachts sailing for 2.5 h.

Distance for Yacht B $= 16 \times 2.5$

$= 40$

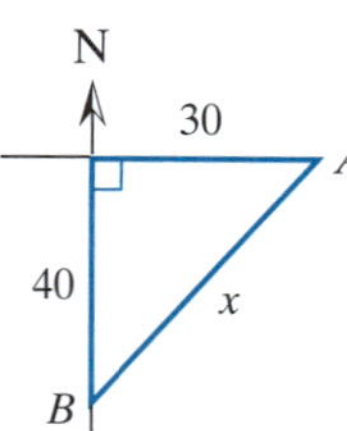

Let $AB = x$.

$$x^2 = 30^2 + 40^2$$
$$= 2500$$
$$x = 50$$

$\therefore$ yachts are 50 km apart.

6. $\frac{\mathbf{25}}{\mathbf{2}}(\boldsymbol{\pi} - \mathbf{2})$ **cm²**

Let radius be x.

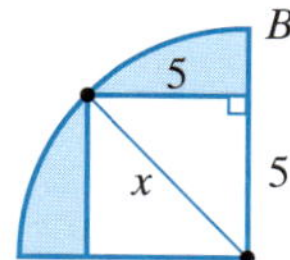

$$x^2 = 5^2 + 5^2$$
$$x^2 = 50$$

$$\text{Area of quadrant} = \frac{1}{4}\pi r^2 = \frac{1}{4}\pi \times x^2 = \frac{1}{4}\pi \times 50 = \frac{25\pi}{2}$$

$$\text{Area of square} = 5^2 = 25$$

Shaded area

= Area quadrant − area square

$= \frac{25\pi}{2} - 25$

$= \frac{25}{2}(\pi - 2)$

∴ the area is $\frac{25}{2}(\pi - 2)$ cm².

Key Skill 30 (pages 76–77)

1. **6(π + 7) cm**

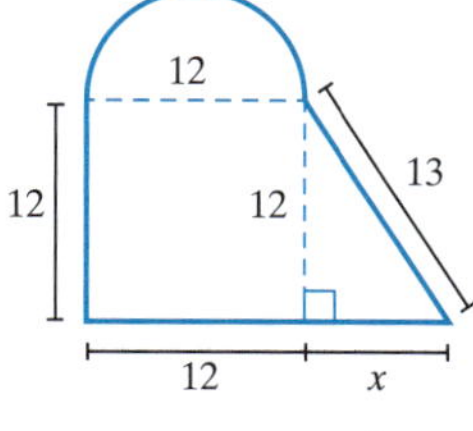

$$x^2 = 13^2 - 12^2 = 25$$
$$x = 5$$

Half circumference

$= \frac{1}{2} \times 2 \times \pi \times 6$

$= 6\pi$

$$\text{Perimeter} = 6\pi + 12 + 17 + 13 = 6\pi + 42 = 6(\pi + 7)$$

∴ the perimeter is $6(\pi + 7)$ cm.

2. **32 cm**

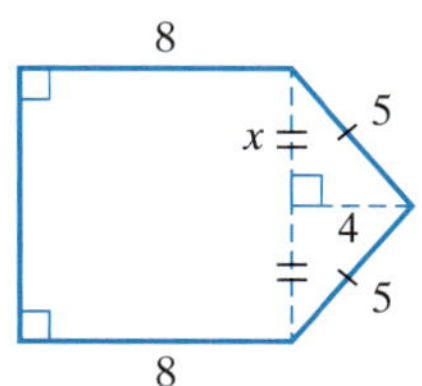

$$x^2 = 5^2 - 4^2 = 9$$
$$x = 3$$
$$\text{Perimeter} = 6 + 8 + 5 + 5 + 8 = 32$$

∴ the perimeter is 32 cm.

3. **2(8 + 5π) cm**

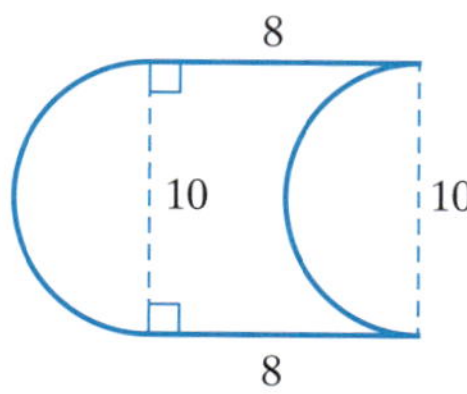

$$\text{Perimeter} = 8 + 8 + 2 \times \pi \times 5 = 16 + 10\pi = 2(8 + 5\pi)$$

∴ the perimeter is $2(8 + 5\pi)$ cm.

4. **(28 + 9π) cm**

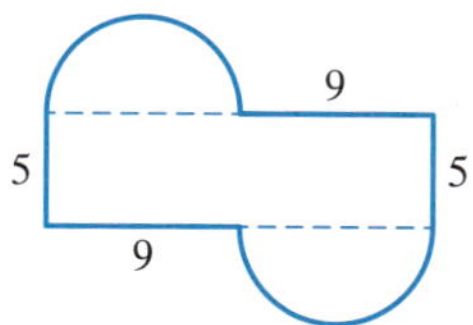

Perimeter

$= 9 + 9 + 5 + 5 + 2 \times \pi \times 4.5$

$= 28 + 9\pi$

∴ the perimeter is $(28 + 9\pi)$ cm.

5. **(12 + √73π) cm**

Let diameter be x.

$$x^2 = 3^2 + 8^2$$
$$x^2 = 73$$
$$x = \sqrt{73}$$

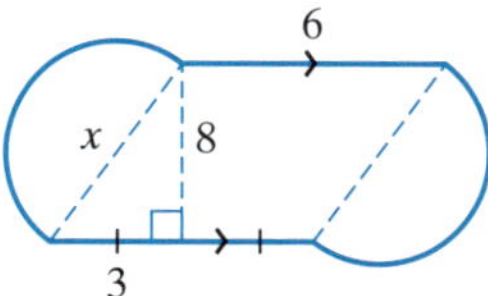

$$\text{Perimeter} = 6 + 6 + 2 \times \pi \times \frac{\sqrt{73}}{2} = 12 + \sqrt{73}\pi$$

∴ the perimeter is $(12 + \sqrt{73}\pi)$ cm.

6. **π cm**

Area of square = 2

Length of side = $\sqrt{2}$

Let radius OD be x.

$$x^2 = (\sqrt{2})^2 + (\sqrt{2})^2$$
$$x^2 = 4$$
$$x = 2$$

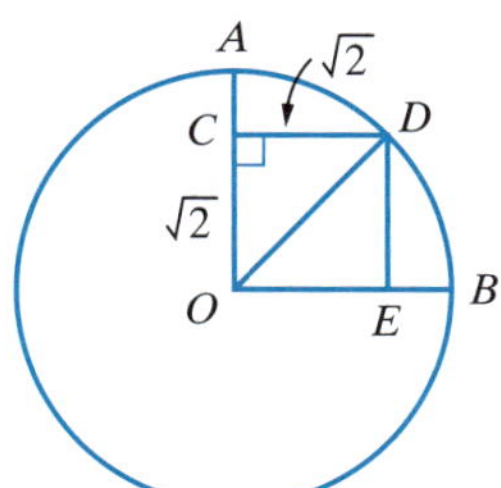

Radius = 2 cm

$$\text{Length of arc} = \frac{1}{4} \times 2 \times \pi \times 2 = \pi$$

∴ the arc has length of π cm.

Key Skill  31 (pages 78–79)

1. **6(5 + 3π) cm²**

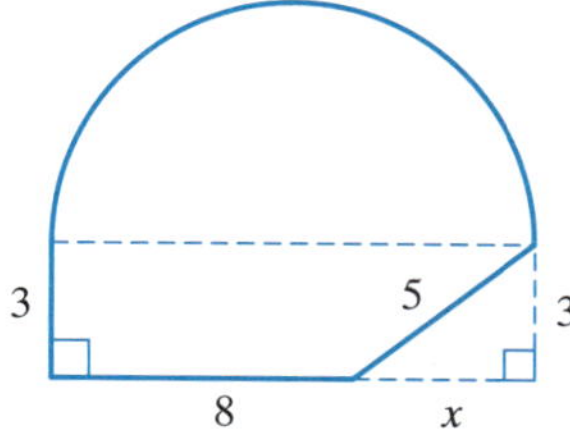

$$x^2 = 5^2 - 3^2 = 16$$
$$x = 4$$

As 8 + 4 = 12, the diameter is 12 cm.

Area = Area of trapezium + area of semicircle

$= \frac{1}{2} \times 3\,(8 + 12) + \frac{1}{2} \times \pi \times 6^2$

$= 30 + 18\pi$

$= 6(5 + 3\pi)$

∴ the area is $6(5 + 3\pi)$ cm².

2. **2π cm²**

Shape = quadrant + 2 semicircles

Let the radius of quadrant be x.

∴ the diameter of the semicircles is x.

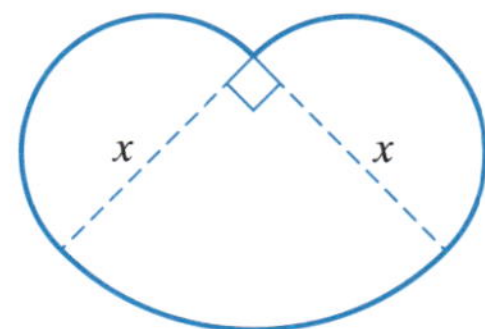

Perimeter

$= \frac{1}{4} \times 2 \times \pi \times x + 2 \times \pi \times \frac{x}{2}$

$= \frac{\pi}{2}x + \pi x$

$= \frac{3\pi}{2}x$

Now, $\frac{3\pi}{2}x = 3\pi$

$x = 2$

$$\text{Area} = \frac{1}{4} \times \pi \times 2^2 + \pi \times 1^2 = 2\pi$$

∴ the area is 2π cm².

3. **16 cm²**

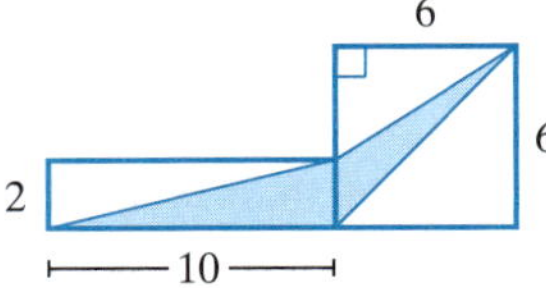

Area $= \frac{1}{2} \times 10 \times 2 + \frac{1}{2} \times 2 \times 6$
$= 16$
$\therefore$ the area is 16 cm².

4. **8π cm²**
Let radius OC be x.

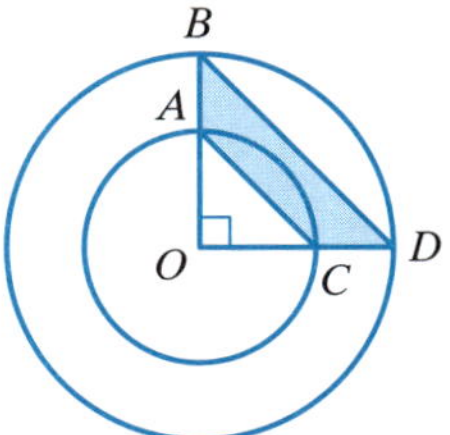

Area $\triangle AOC = \frac{1}{2} \times x \times x$
$= \frac{x^2}{2}$
Also, the radius $OD = 2x$
Area $\triangle BOD = \frac{1}{2} \times 2x \times 2x$
$= 2x^2$
Shaded area $= 2x^2 - \frac{x^2}{2}$
$= \frac{3x^2}{2}$
$\therefore \frac{3x^2}{2} = 12$
$3x^2 = 24$
$x^2 = 8$
Area $= \pi \times x^2$
$= 8\pi$
$\therefore$ the area is 8π cm².

5. **36π cm²**

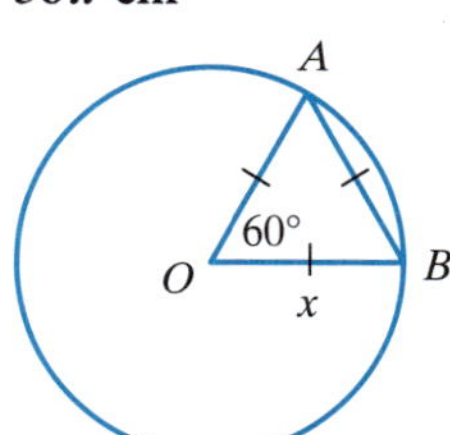

As $AO = OB = AB$, then $\triangle AOB$ is equilateral.
$\therefore \angle AOB = 60°$.
Let radius OB be x.
Length of arc $AB = \frac{1}{6} \times 2 \times \pi \times x$
$= \frac{\pi x}{3}$
$\therefore \frac{\pi x}{3} = 2\pi$
$\pi x = 6\pi$
$x = 6$
$\therefore$ the radius is 6 cm.
Area $= \pi \times 6^2$
$= 36\pi$
$\therefore$ the area is 36π cm².

6. **$\frac{1}{\pi}$**
In square $OCDE$, let $OE = DE = y$, and $OD = x$.

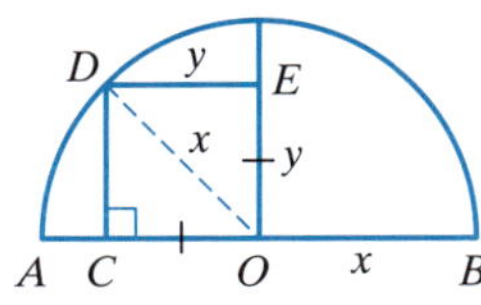

$x^2 = y^2 + y^2$
$= 2y^2$
But area of square $= y^2 = \frac{x^2}{2}$
Also, area of semicircle
$= \frac{1}{2} \times \pi \times x^2 = \frac{\pi x^2}{2}$
Fraction $= \frac{x^2}{2} \div \frac{\pi x^2}{2}$
$= \frac{x^2}{2} \times \frac{2}{\pi x^2}$
$= \frac{1}{\pi}$
$\therefore$ the fraction is $\frac{1}{\pi}$.

Key Skill  32 (pages 80–81)

1. **896 cm²**

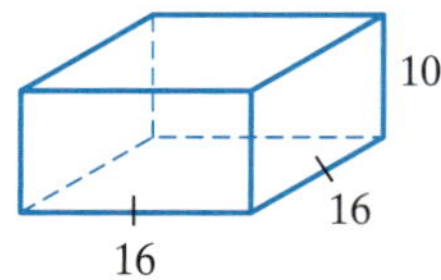

$SA = 16^2 + 4 \times 16 \times 10$
$= 896$
$\therefore$ the surface area is 896 cm².

2. **660 cm²**

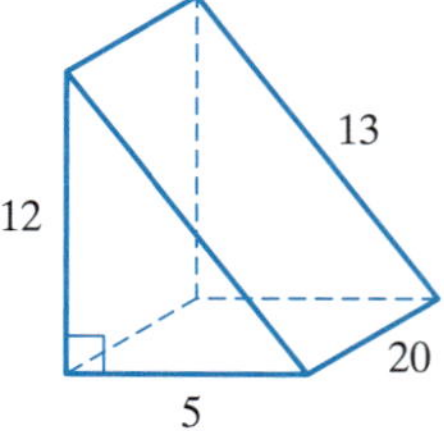

As {3, 4, 5} is a Pythagorean triad, then the triangle is right-angled.
$SA = 2 \times \frac{1}{2} \times 5 \times 12 + 20 \times 5 + 13 \times 20 + 12 \times 20$
$= 660$
$\therefore$ the surface area is 660 cm².

3. **22.4 m²**

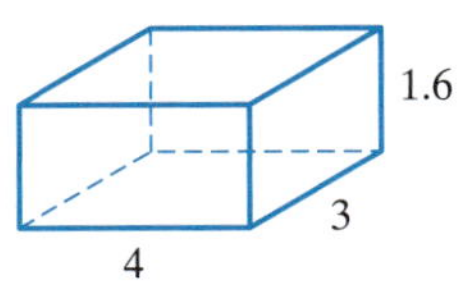

$SA = 2 \times 4 \times 1.6 + 2 \times 3 \times 1.6$
$= 22.4$
$\therefore$ the surface area is 22.4 m².

4. **8 cm**
Let the height be x.

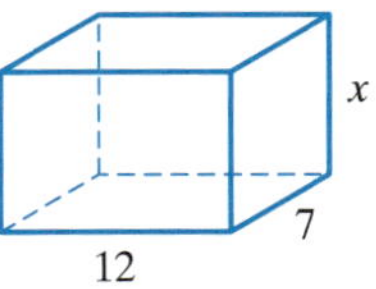

$2 \times 12 \times 7 + 2 \times 12 \times x + 2 \times 7 \times x = 472$
$168 + 24x + 14x = 472$
$38x = 472 - 168$
$38x = 304$
$x = 8$
$\therefore$ the height is 8 cm.

5. **40.6 L**
Let the hypotenuse be x.

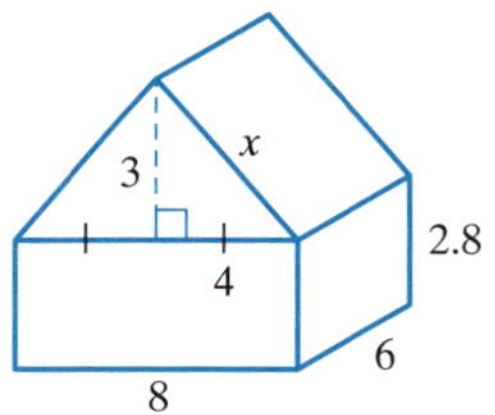

$x^2 = 3^2 + 4^2$
$= 9 + 16$
$= 25$
$x = 5$
$SA = 2 \times 5 \times 6 + 2 \times 8 \times 2.8 + 2 \times 6 \times 2.8 + 2 \times \frac{1}{2} \times 3 \times 8$
$= 162.4$
Area of two coats $= 162.4 \times 2$
$= 324.8$
$\therefore$ the area is 324.8 m²
Number of litres $= 324.8 \div 8$
$= 40.6$
$\therefore$ will need 40.6 L.

6. 6 cm

Let the side of the base be x.

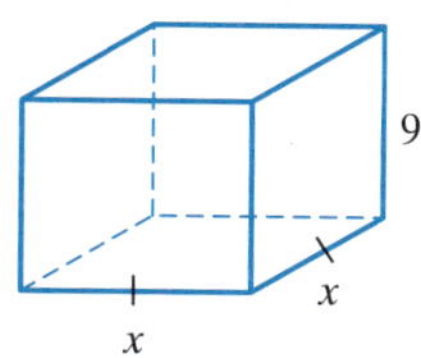

$$2x^2 + 4 \times 9 \times x = 288$$
$$2x^2 + 36x - 288 = 0$$
$$x^2 + 18x - 144 = 0$$
$$(x + 24)(x - 6) = 0$$
$$x = -24, 6$$
$$x = 6 \quad (\text{as } x > 0)$$

$\therefore$ the length is 6 cm.

Key Skill 33 (pages 82–83)

1. \$4698

$$\begin{aligned} SA &= 2\pi r^2 + 2\pi rh \\ &= 2 \times \pi \times 3.2^2 \\ &\quad + 2 \times \pi \times 3.2 \times 9.5 \\ &= 255.348\,6509\ldots \\ &= 255.35 \text{ (2 dec. pl.)} \\ \text{Cost} &= 255.35 \times 18.4 \\ &= 4698.44 \\ &= 4698 \text{ (nearest whole)} \end{aligned}$$

$\therefore$ the cost was \$4698.

2. 13 239 cm^2

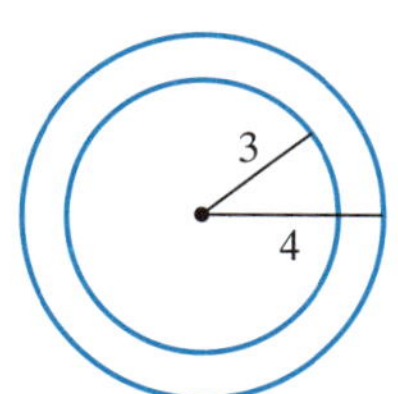

3 m = 300 cm

Outer radius = 4 cm

Inner radius = 3 cm

$$\begin{aligned} \text{Outer SA} &= 2\pi rh \\ &= 2 \times \pi \times 4 \times 300 \\ &= 7539.822\,369\ldots \\ \text{Inner SA} &= 2\pi rh \\ &= 2 \times \pi \times 3 \times 300 \\ &= 5654.866\,776\ldots \end{aligned}$$

SA of 2 ends

$= 2 \times (\pi \times 4^2 - \pi \times 3^2)$

$= 43.982\,297\,15$

$$\begin{aligned} \text{Total SA} &= 13\,238.671\,44\ldots \\ &= 13\,239 \text{ (nearest whole)} \end{aligned}$$

$\therefore$ the total surface area is 13 239 cm^2.

3. 226 m^2

Radius = 0.4, height = 0.9

$$\begin{aligned} \text{Curved SA} &= 2\pi rh \\ &= 2 \times \pi \times 0.4 \times 0.9 \\ &= 2.261\,946\,711\ldots \\ &= 2.262 \text{ (3 dec. pl.)} \\ \text{Area rolled} &= 100 \times 2.262 \\ &= 226 \text{ (nearest whole)} \end{aligned}$$

$\therefore$ the area rolled is 226 m^2.

4. 247 cm^2

Radius = 3.75, height = 10.5

$$\begin{aligned} \text{Curved SA} &= 2\pi rh \\ &= 2 \times \pi \times 3.75 \times 10.5 \\ &= 247.400\,4215\ldots \\ &= 247 \text{ (3 sig. figs.)} \end{aligned}$$

$\therefore$ the area of the label is 247 cm^2.

5. \$24 000

$$\begin{aligned} \text{Curved SA} &= \frac{1}{4} \times 2\pi rh \\ &= \frac{1}{4} \times 2 \times \pi \times 12 \times 32 \\ &= 192\pi \end{aligned}$$

Total SA

$= 192\pi + 2 \times \frac{1}{4} \times \pi \times 12^2$
$\quad + 2 \times 12 \times 32$

$= 192\pi + 72\pi + 768$

$= 1597.380\,461\ldots$

$= 1597.38$ (2 dec. pl.)

$$\begin{aligned} \text{Cost} &= 1597.38 \times 15 \\ &= 23\,960.706\,91\ldots \\ &= 24\,000 \text{ (nearest thousand)} \end{aligned}$$

$\therefore$ the cost is \$24 000.

6. 448π cm^2.

Heights of cylinders are 14 cm and 8 cm.

Total of curved SA

$= 2 \times \pi \times 6 \times 8 + 2 \times \pi \times 8 \times 14$

$= 320\pi$

Total of horizontal faces

$= \pi \times 6^2 + \pi \times 8^2$
$\quad + (\pi \times 8^2 - \pi \times 6^2)$

$= 128\pi$

$$\begin{aligned} \text{Total SA} &= 320\pi + 128\pi \\ &= 448\pi \end{aligned}$$

$\therefore$ the surface area is 448π cm^2.

Key Skill 34 (pages 84–85)

1. 216π cm^2

Let the radius be x.

$$\begin{aligned} CSA &= 2\pi rh \\ 144\pi &= 2 \times \pi \times x \times 12 \\ 24\pi x &= 144\pi \\ x &= 6 \end{aligned}$$

$\therefore$ the radius is 6 cm.

$$\begin{aligned} SA &= 2\pi r^2 + 2\pi rh \\ &= 2 \times \pi \times 6^2 + 144\pi \\ &= 72\pi + 144\pi \\ &= 216\pi \end{aligned}$$

$\therefore$ the total surface area is 216π cm^2.

2. \$4725

Let the radius be x.

$$\begin{aligned} SA &= 2\pi r^2 + 2\pi rh \\ 288\pi &= 2 \times \pi \times x^2 \\ &\quad + 2 \times \pi \times x \times 10 \\ 2\pi x^2 + 20\pi x - 288\pi &= 0 \\ x^2 + 10x - 144 &= 0 \\ (x + 18)(x - 8) &= 0 \\ x &= -18, 8 \\ x &= 8 \quad (\text{as } x > 0) \\ CSA &= 2 \times \pi \times 8 \times 10 \\ &= 160\pi \\ \text{Cost} &= 160\pi \times 9.4 \\ &= 4724.955\,351\ldots \\ &= 4725 \text{ (nearest whole)} \end{aligned}$$

$\therefore$ the cost is \$4725.

3. 16 cm

Let the height be x.

$$\begin{aligned} SA &= 2\pi r^2 + 2\pi rh \\ 450\pi &= 2 \times \pi \times 9^2 \\ &\quad + 2 \times \pi \times 9 \times x \\ 162\pi + 18\pi x &= 450\pi \\ 18\pi x &= 288\pi \\ x &= 16 \end{aligned}$$

$\therefore$ the height is 16 cm.

4. 300 cm^2

Let the radius be x.

$$\begin{aligned} SA &= 2\pi r^2 + 2\pi rh \\ 128\pi &= 2 \times \pi \times x^2 \\ &\quad + 2 \times \pi \times x \times 12 \\ 2\pi x^2 + 24\pi x - 128\pi &= 0 \\ x^2 + 12x - 64 &= 0 \\ (x + 16)(x - 4) &= 0 \\ x &= -16, 4 \\ x &= 4 \quad (\text{as } x > 0) \\ CSA &= 2 \times \pi \times 4 \times 12 \\ &= 96\pi \\ &= 301.592\,8947\ldots \\ &= 300 \text{ (correct to 2 sig. figs)} \end{aligned}$$

$\therefore$ the curved surface area is 300 cm^2.

5. 201.1 cm^2

Let the radius be x.

$$\begin{aligned} CSA &= 2\pi rh \\ 160\pi &= 2 \times \pi \times x \times 10 \\ 20\pi x &= 160\pi \\ x &= 8 \end{aligned}$$

$\therefore$ the radius is 8 cm.

$$\begin{aligned} A &= \pi \times 8^2 \\ &= 201.061\,9298\ldots \\ &= 201.1 \text{ (1 dec. pl.)} \end{aligned}$$

$\therefore$ the shaded area is 201.1 cm^2.

6. 8 cm, 4 cm

Let the inner radius be x.

$$\begin{aligned}
2 \times \pi \times 10 \times 12 & \\
+ 2 \times \pi \times x \times 12 & \\
+ 2 \times \pi \times (10^2 - x^2) &= 504\pi \\
240\pi + 24\pi x + 2\pi(100 - x^2) & \\
&= 504\pi \\
240 + 24x + 200 - 2x^2 &= 504 \\
2x^2 - 24x + 64 &= 0 \\
x^2 - 12x + 32 &= 0 \\
(x - 8)(x - 4) &= 0 \\
x &= 8, 4
\end{aligned}$$

$\therefore$ the radii are 8 cm and 4 cm.

Revision Test  (page 86)

1. $1622

Let the two lengths be x and y.

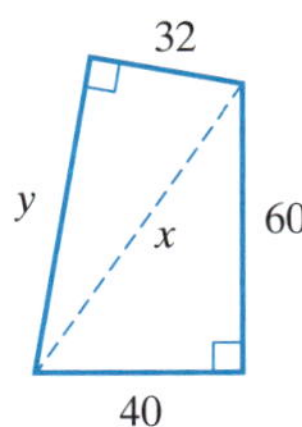

$$\begin{aligned}
x^2 &= 40^2 + 60^2 \\
&= 5200 \\
y^2 &= x^2 - 32^2 \\
&= 5200 - 1024 \\
&= 4176 \\
y &= \sqrt{4176} \\
&= 64.621\,977\,69\ldots \\
&= 64.622 \text{ (3 dec. pl.)} \\
\text{Perimeter} &= 40 + 60 + 32 + 64.622 \\
&= 196.622 \\
\text{Cost} &= 8.25 \times 196.622 \\
&= 1622.1315 \\
&= 1622 \text{ (nearest whole)}
\end{aligned}$$

$\therefore$ the cost is $1622.

2. 2200 km

Let the distance be x.

Use Distance = Speed × Time

Jet A: $1000 \times 2 = 2000$

Jet B: $1200 \times 0.75 = 900$

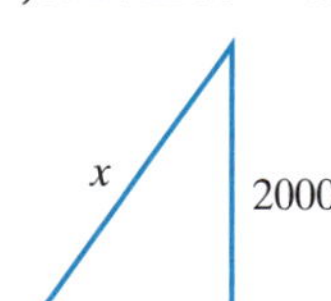

$$\begin{aligned}
x^2 &= 900^2 + 2000^2 \\
&= 4\,810\,000 \\
x &= 2193.171\,22\ldots \\
&= 2200 \text{ (nearest hundred)}
\end{aligned}$$

$\therefore$ the jets are 2200 km apart.

3. 9.25 cm

Let the width be x cm.

$$\begin{aligned}
2 \times 10 \times x + 2 \times 6 \times x & \\
+ 2 \times 10 \times 6 &= 416 \\
32x + 120 &= 416 \\
32x &= 296 \\
x &= \frac{296}{32} \\
&= 9.25
\end{aligned}$$

$\therefore$ the width is 9.25 cm.

4. 396 cm^3

Let EC, FC be x cm.

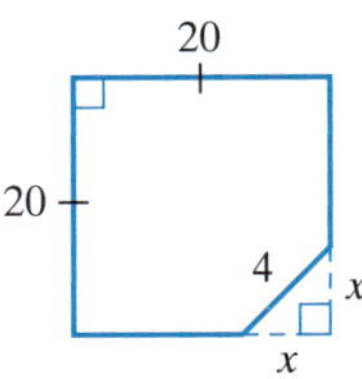

$$\begin{aligned}
x^2 + x^2 &= 4^2 \\
2x^2 &= 16 \\
x^2 &= 8 \\
\text{Area} &= 20^2 - \frac{1}{2} \times x \times x \\
&= 400 - \frac{1}{2} \times x^2 \\
&= 400 - 4 \\
&= 396
\end{aligned}$$

$\therefore$ the area is 396 cm^2.

5. $(\frac{161\pi}{2} + 128)$ cm^2

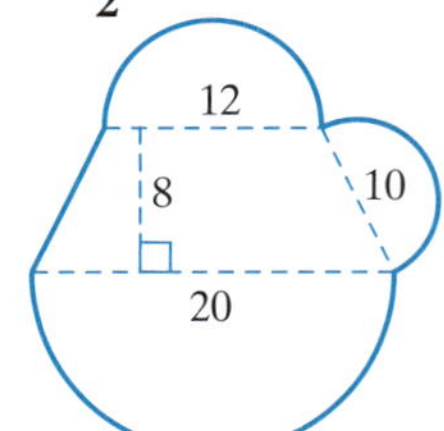

$$\begin{aligned}
\text{Area} &= \frac{1}{2} \times \pi \times 6^2 + \frac{1}{2} \times \pi \times 5^2 \\
&+ \frac{1}{2} \times \pi \times 10^2 + \frac{1}{2} \times 8(12 + 20) \\
&= \frac{161\pi}{2} + 128
\end{aligned}$$

$\therefore$ the area is $(\frac{161\pi}{2} + 128)$ cm^2.

6. 990π cm^3

$$\begin{aligned}
\text{Curved SA} &= 2\pi rh \\
540\pi &= 2 \times \pi \times r \times 18 \\
36\pi r &= 540\pi \\
r &= \frac{540\pi}{36\pi} \\
&= 15
\end{aligned}$$

$\therefore$ the radius is 15 cm.

$$\begin{aligned}
\text{SA} &= 2\pi r^2 + 2\pi rh \\
&= 2 \times \pi \times 15^2 + 540\pi \\
&= 990\pi
\end{aligned}$$

$\therefore$ the total surface area is 990π cm^2.

Revision Test  (page 87)

1. $\pi : 4$

Let the pieces of string be 4 m long each.

$$\begin{aligned}
\text{Square:} \quad \text{Side} &= 1 \text{ m} \\
\text{Area} &= 1 \times 1 \\
&= 1
\end{aligned}$$

$\therefore$ the area of square is 1 m^2.

$$\begin{aligned}
\text{Circle:} \quad C &= 2\pi r \\
4 &= 2\pi r \\
r &= \frac{4}{2\pi} \\
&= \frac{2}{\pi} \\
A &= \pi r^2 \\
&= \pi \times \left(\frac{2}{\pi}\right)^2 \\
&= \frac{4}{\pi}
\end{aligned}$$

$\text{Ratio} = 1 : \frac{4}{\pi}$, or $\pi : 4$.

$\therefore$ the ratio is $\pi : 4$.

2. $\frac{\pi\sqrt{8}}{4}$ cm

Let the radius be x.

As area of square is 4 cm^2, then side = 2 cm.

$\angle SOQ = 45°$.

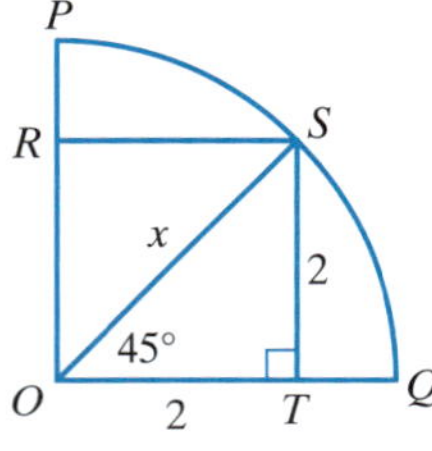

$$\begin{aligned}
x^2 &= 2^2 + 2^2 \\
&= 8 \\
x &= \sqrt{8} \\
\text{Length of arc } SQ &= \frac{45}{360} \times 2\pi r \\
&= \frac{1}{8} \times 2 \times \pi \times \sqrt{8} \\
&= \frac{\pi\sqrt{8}}{4}
\end{aligned}$$

$\therefore$ the length of the arc is $\frac{\pi\sqrt{8}}{4}$ cm.

3. 18 : 25

$$\begin{aligned}
\text{Vol. of large cube} &= 3^3 + 4^3 + 5^3 \\
&= 216
\end{aligned}$$

$\therefore$ the volume of the large cube is 216 cm^3.

As $\sqrt[3]{216} = 6$, sides of large cube are 6 cm.

$$\begin{aligned}
\text{SA of large cube} &= 6 \times 6^2 \\
&= 216
\end{aligned}$$

Total SA of other cubes
$= 6 \times 3^2 + 6 \times 4^2 + 6 \times 5^2$
$= 300$
Ratio $= 216:300$
$= 18:25$
$\therefore$ the ratio is 18 : 25.

4. $(4 + \sqrt{32})$ mm

Form a square by joining centres of small circle.
Let the hypotenuse be x.
$x^2 = 4^2 + 4^2$
$= 32$
$x = \sqrt{32}$
Diameter of large circle
$= \sqrt{32} + 2 \times 2$
$= 4 + \sqrt{32}$
$\therefore$ the diameter is $(4 + \sqrt{32})$ mm.

5. 1500 cm³
Let the side of the square base be x cm.
$\text{SA} = 2 \times x^2 + 4 \times 15 \times x$
$800 = 2x^2 + 60x$
$2x^2 + 60x - 800 = 0$
$x^2 + 30x - 400 = 0$
$(x + 40)(x - 10) = 0$
$x = -40, 10$
$x = 10 \ (x > 0)$
$\therefore$ the dimensions are 10 cm, 10 cm, 15 cm.
Volume $= 10 \times 10 \times 15$
$= 1500$
$\therefore$ the volume is 1500 cm³.

6. $(464 + 72\pi)$ cm²
$\text{SA} = 2 \times 8^2 + 3 \times 14 \times 8 + \pi \times 4^2 + \frac{1}{2} \times 2 \times \pi \times 4 \times 14$
$= 128 + 336 + 16\pi + 56\pi$
$= 464 + 72\pi$
$\therefore$ the surface area is $(464 + 72\pi)$ cm².

Key Skill 35 (pages 88–89)

1. 68.8 kg
$A = \frac{1}{2}h(a + b)$
$= \frac{1}{2} \times 20(15 + 28)$
$= 430$
$V = Ah$
$= 430 \times 10$
$= 4300$
$\therefore$ the volume is 4300 cm³.
Mass in g $= 4300 \times 16$
$= 68\,800$
Mass in kg $= 68\,800 \div 1000$
$= 68.8$
$\therefore$ the mass is 68.8 kg.

2. 240 cm³
Let the height be x.

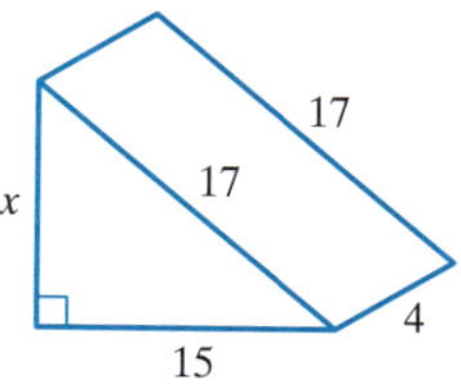

$x^2 = 17^2 - 15^2$
$= 64$
$x = 8$
$V = \frac{1}{2} \times 15 \times 8 \times 4$
$= 240$
$\therefore$ the volume is 240 cm³.

3. 32 cm
SA of cube $= 6s^2$
$6s^2 = 3456$
$s^2 = 576$
$s = 24 \quad (s > 0)$
Vol. of sand $= 24 \times 24 \times 12$
$= 6912$
Vol. of prism $= lbh$
$6912 = 18 \times 12 \times h$
$216h = 6912$
$h = 32$
$\therefore$ the sand is 32 cm deep.

4. 72 cm²
Let the width be x.
$\therefore$ the length is $(x + 3)$.
$V = lbh$
$36 = (x + 3) \times x \times 2$
$2x(x + 3) = 36$
$2x^2 + 6x - 36 = 0$
$x^2 + 3x - 18 = 0$
$(x + 6)(x - 3) = 0$
$x = -6, 3$
$x = 3 \quad (x > 0)$
$\therefore$ the dimensions are 6 cm, 3 cm, 2 cm.
$\text{SA} = 2 \times 6 \times 3 + 2 \times 6 \times 2 + 2 \times 3 \times 2$
$= 72$
$\therefore$ the surface area is 72 cm².

5. 75%
Let the base of small triangle be x.

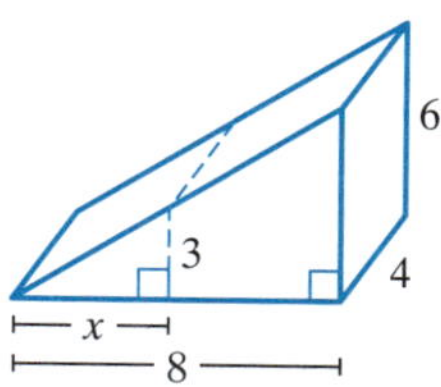

Using similar triangles (triangles are equiangular):
$\frac{x}{8} = \frac{3}{6}$ (matching sides in proportion)
$x = 4$ (see Key Skill 39)
$\therefore$ 4 cm of wedge under door, 4 cm of wedge seen.
Volume of entire wedge
$= \frac{1}{2} \times 8 \times 6 \times 4$
$= 96$
Volume of unseen wedge
$= \frac{1}{2} \times 4 \times 3 \times 4$
$= 24$
Percentage unseen $= \frac{24}{96} \times 100\%$
$= 25\%$
As $110 - 25 = 75$, then 75% seen.
$\therefore$ 75% of the volume of the wedge is visible.

6. 2980 cm²
Let the height of triangle be x.

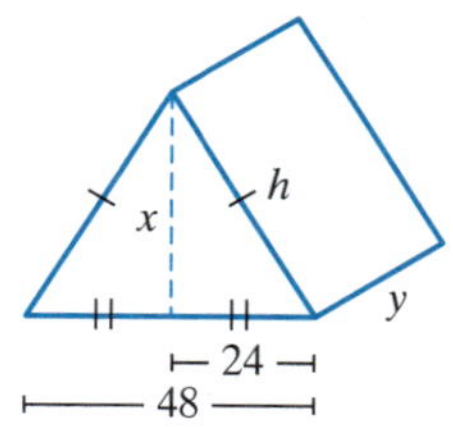

Area $= \frac{1}{2}bh$
$240 = \frac{1}{2} \times 48 \times x$
$24x = 240$
$x = 10$
Let the 'height' of the prism be y.
Using the volume of 6000:
$240y = 6000$
$y = 25$
Also, let the hypotenuse be h.
$h^2 = 10^2 + 24^2$
$= 676$
$h = 26$
$\text{SA} = 2 \times 240 + 48 \times 25 + 2 \times 26 \times 25$
$= 2980$
$\therefore$ the surface area is 2980 cm².

Key Skill 36 (pages 90–91)

1. 1.16 m^3

Radius = 40 cm = 0.4 m

$$V = \pi \times 0.4^2 \times 2.3$$
$$= 1.156\,106\,097\ldots$$
$$= 1.16 \text{ (2 dec. pl.)}$$

$\therefore$ the volume is 1.16 m^3.

2. 0.07 m^3

Radius = 15 cm = 0.15 m

$$V = \frac{1}{2} \times \pi \times 0.15^2 \times 2.1$$
$$= 0.074\,220\,126\ldots$$
$$= 0.07 \text{ (2 dec. pl.)}$$

$\therefore$ the volume is 0.07 m^3.

3. 1.31 m^3

$$V = \frac{1}{4} \times \pi \times 0.7^2 \times 3.4$$
$$= 1.308\,473\,34\ldots$$
$$= 1.31 \text{ (2 dec. pl.)}$$

$\therefore$ the volume is 1.31 m^3.

4. 343.36 cm^3

$$V = 10 \times 10 \times 16 - \frac{1}{4} \times \pi \times 10^2 \times 16$$
$$= 343.362\,9386\ldots$$
$$= 343.36 \text{ (2 dec. pl.)}$$

$\therefore$ the volume is 343.36 cm^3.

5. 3811.96 cm^3

$$V = (12 \times 12 + \frac{75}{360} \times \pi \times 12^2) \times 16$$
$$= 3811.964\,474\ldots$$
$$= 3811.96 \text{ (2 dec. pl.)}$$

$\therefore$ the volume is 3811.96 cm^3.

6. 39 kg

Radii are 12 cm and 4 cm:

$$V = [\pi(12^2 - 4^2)] \times 30$$
$$= 12\,063.715\,79\ldots$$
$$= 12\,063.72 \text{ (2 dec. pl.)}$$
$$\text{Mass} = 12\,063.72 \times 3.2$$
$$= 38\,603.904\ldots$$
$$= 38\,604 \text{ (nearest whole)}$$
$$\text{Mass in kg} = 38\,604 \div 1000$$
$$= 38.604$$
$$= 39 \text{ (nearest whole)}$$

$\therefore$ the mass is 39 kg.

Key Skill 37 (pages 92–93)

1. $\frac{1458}{\pi}$ cm^3

$$C = 2\pi r$$
$$18 = 2\pi r$$
$$\therefore \quad r = \frac{18}{2\pi}$$
$$= \frac{9}{\pi}$$
$$V = \pi r^2 h$$
$$= \pi \times \left(\frac{9}{\pi}\right)^2 \times 18$$
$$= \frac{1458}{\pi}$$

$\therefore$ the volume is $\frac{1458}{\pi}$ cm^3.

2. 2000π cm^3

Let the radius be x.

$$SA = 2\pi r^2 + 2\pi rh$$
$$600\pi = 2 \times \pi \times x^2 + 2 \times \pi \times x \times 20$$
$$2\pi x^2 + 40\pi x - 600\pi = 0$$
$$x^2 + 20x - 300 = 0$$
$$(x + 30)(x - 10) = 0$$
$$x = -30, 10$$
$$x = 10 \quad (x > 0)$$

$\therefore$ the radius is 10 cm.

$$V = \pi r^2 h$$
$$= \pi \times 10^2 \times 20$$
$$= 2000\pi$$

$\therefore$ the volume is 2000π cm^3.

3. $\frac{5184}{\pi}$ cm^3

$$C = 2\pi r$$
$$48 = 2\pi r$$
$$r = \frac{48}{2\pi}$$
$$= \frac{24}{\pi}$$
$$CSA = 2\pi rh$$
$$432 = 2 \times \pi \times \frac{24}{\pi} \times h$$
$$432 = 48h$$
$$h = 9$$

$\therefore$ the radius is $\frac{24}{\pi}$ cm, height is 9 cm.

$$V = \pi r^2 h$$
$$= \pi \times \left(\frac{24}{\pi}\right)^2 \times 9$$
$$= \frac{5184}{\pi}$$

$\therefore$ the volume is $\frac{5184}{\pi}$ cm^3.

4. 203.72 cm^3

$$CSA = 2\pi rh$$
$$160 = 2 \times \pi \times r \times 10$$
$$20\pi r = 160$$
$$r = \frac{160}{20\pi}$$
$$= \frac{8}{\pi}$$
$$V = \pi r^2 h$$
$$= \pi \times \left(\frac{8}{\pi}\right)^2 \times 10$$
$$= \frac{640}{\pi}$$
$$= 203.72 \text{ (2 dec. pl.)}$$

$\therefore$ the volume is 203.72 cm^3.

5. 253.08 cm^3

Let the hypotenuse be x.

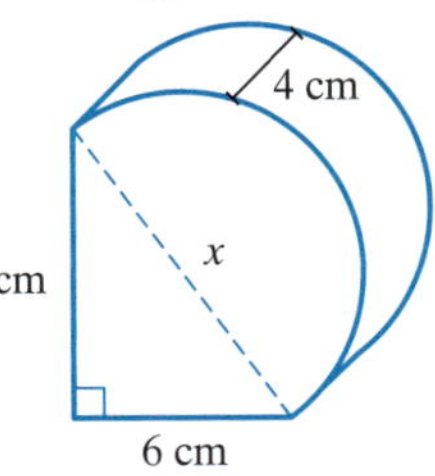

$$x^2 = 6^2 + 8^2$$
$$= 100$$
$$x = 10$$
$$\text{Vol} = \frac{1}{2} \times 6 \times 8 \times 4 + \frac{1}{2} \times \pi \times 5^2 \times 4$$
$$= 96 + 50\pi$$
$$= 253.079\,6327\ldots$$
$$= 253.08 \text{ (2 dec. pl.)}$$

$\therefore$ the volume is 253.08 cm^3.

6. 160(2π + 5) cm^3

Let the side be x.

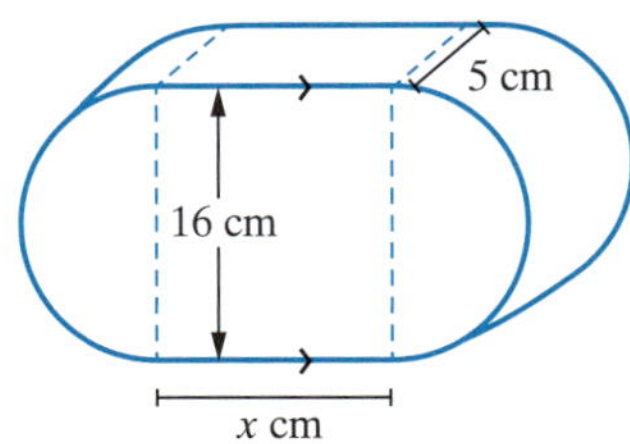

Radius of semicircles is 8 cm.

$$SA = 2 \times \pi \times 8 \times 5 + 2 \times \pi \times 8^2 + 2 \times 16 \times x + 2 \times x \times 5$$
$$4(52\pi + 105) = 80\pi + 128\pi + 32x + 10x$$
$$208\pi + 420 = 208\pi + 42x$$
$$42x = 420$$
$$x = 10$$
$$\text{Vol.} = [16 \times 10 + \pi \times 8^2] \times 5$$
$$= [160 + 64\pi] \times 5$$
$$= 5(64\pi + 160)$$
$$= 160(2\pi + 5)$$

$\therefore$ the volume is 160(2π + 5) cm^3.

Key Skill 38 (pages 94–95)

1. 36 L

Radius 20 cm.

$$V = \frac{1}{3} \times \pi r^2 h$$
$$= \frac{1}{3} \times \pi \times 20^2 \times 85$$
$$= 35\,604.716\,74\ldots$$
$$= 35\,605 \text{ (nearest whole)}$$

$\therefore$ volume is 35 605 cm^3

Capacity is 35 605 mL, or 36 L (nearest whole).

$\therefore$ contains 36 litres.

2. **1 h 42 m 40 s**

$V = lbh$
$= 1.4 \times 0.8 \times 1.1$
$= 1.232$

$\therefore$ volume is 1.232 m^3

Capacity is 1232 L.

As 1L/5 sec, then 12 L/min:

No. of minutes $= 1232 \div 12$
$= 102.666\,6667\ldots$

No. of hours $= 102.666\,6667\ldots \div 60$
$= 1.711\,111\,11\ldots$
$= 1$ h 42 m 40 s

$\therefore$ it will take 1 h 42 m 40 s.

3. **61 cm**

Vol. of cube $= 25^3$
$= 15\,625$

$\therefore$ volume is 15 625 cm^3

Vol. of cylinder $= \pi r^2 h$
$15\,625 = \pi \times 9^2 \times h$
$81\pi h = 15\,625$
$h = \frac{15\,625}{81\pi}$
$= 61.402\,370\,02\ldots$
$= 61$ (nearest whole)

$\therefore$ the height is 61 cm.

4. **43.5%**

2 litres $= 2000$ mL $= 2000$ cm^3

$V = \pi r^2 h$
$= \pi \times 6^2 \times 10$
$= 1130.973\,355\ldots$
$= 1130.97$ (2 dec. pl.)

Water remaining $= 2000 - 1130.97$
$= 869.03$

% remaining $= \frac{869.03}{2000} \times 100\%$
$= 43.4515$
$= 43.5$ (1 dec. pl.)

$\therefore$ 43.5% of water is left in the container.

5. **22 cm**

$V = \pi r^2 h$
$= \frac{3}{4} \times \pi \times 12^2 \times 16$
$= 5428.672\,105\ldots$
$= 5428.67$ (2 dec. pl.)

$\therefore$ volume is 5428.67 cm^3

Let side of cube be x.

$V = \frac{1}{2} \times x^3$
$5428.67 = \frac{1}{2} \times x^3$
$x^3 = 10\,857.344\,21\ldots$
$x = 22.143\,241\,78\ldots$
$= 22$ (nearest whole)

$\therefore$ each side is 22 cm.

6. **6**

$V = \pi r^2 h$
$= \pi \times 4^2 \times 12$
$= 603.185\,7895\ldots$
$= 603.19$ (2 dec. pl.)

Space in cylinder $= \frac{1}{4} \times 603.19$
$= 150.796\,4474\ldots$
$= 150.80$ (2 dec. pl.)

Vol. of each cube $= 4 \times 3 \times 2$
$= 24$

Number of cubes $= 150.80 \div 24$
$= 6.283\,185\,307\ldots$

$\therefore$ Jack can use 6 ice-blocks.

Key Skill  (pages 96–97)

1. **10.67 m**

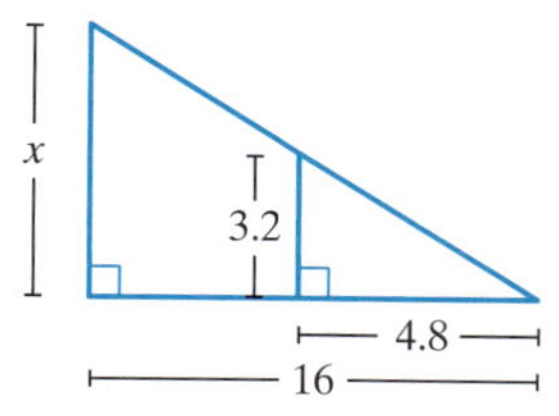

Let the height of the tree be x.

$\frac{x}{3.2} = \frac{16}{4.8}$ (matching sides in sim. $\triangle$s in prop.)

$x = \frac{16 \times 3.2}{4.8}$
$= 10.6666666\ldots$
$= 10.67$ (2 dec. pl.)

$\therefore$ the height is 10.67 m.

2. **5 cm**

Let BC be x.

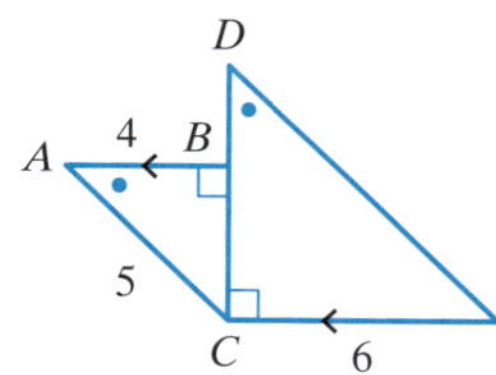

By Pythag: $x^2 = 5^2 - 4^2$
$= 9$
$x = 3$

Let CD be y.

$\frac{y}{4} = \frac{6}{3}$ (matching sides in sim. $\triangle$s in prop)

$y = \frac{4 \times 6}{3}$
$= 8$

$DB = 8 - 3$
$= 5$

$\therefore$ the length of DB is 5 cm.

3. **40.375 m**

Let the length of the flagpole be x.

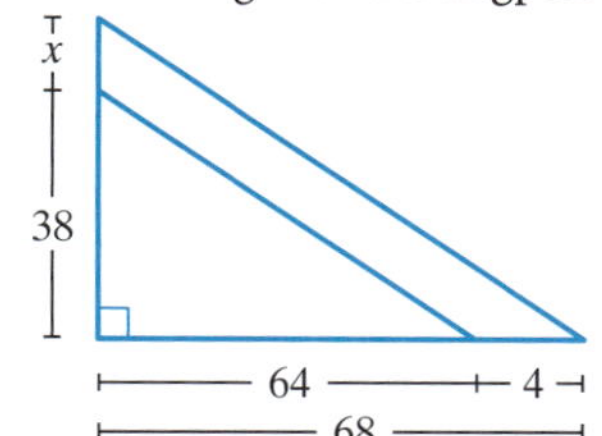

$\frac{x + 38}{38} = \frac{68}{64}$ (matching sides in sim. $\triangle$s in prop.)

$x + 38 = \frac{38 \times 68}{64}$
$x + 38 = 40.375$
$x = 2.375$

$\therefore$ the flagpole is 2.375 m.

$x + 38 = 2.375 + 38$
$= 40.375$

$\therefore$ The top of the flagpole is 40.375 m above the ground.

4. **$33\frac{1}{3}$**

Let $AC = x$ and $DC = y$.

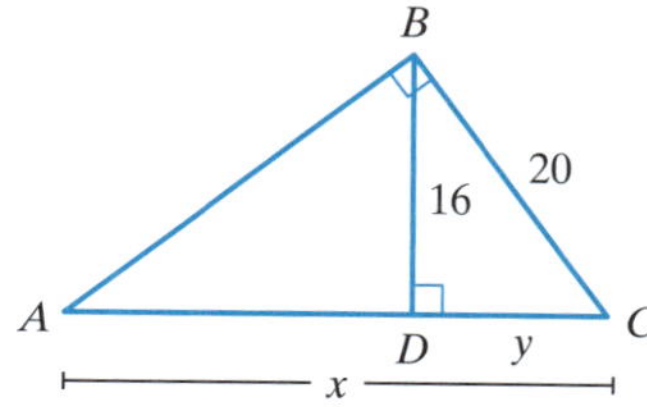

By Pythag: $y^2 = 20^2 - 16^2$
$= 144$
$y = 12$

$\frac{x}{20} = \frac{20}{12}$ (matching sides in sim. $\triangle$s in prop.)

$x = \frac{20 \times 20}{12}$
$= 33\frac{1}{3}$

$\therefore$ AC is $33\frac{1}{3}$.

5. **390 937.5 km**

Let the distance to the moon be x.

Convert dimensions to metres.

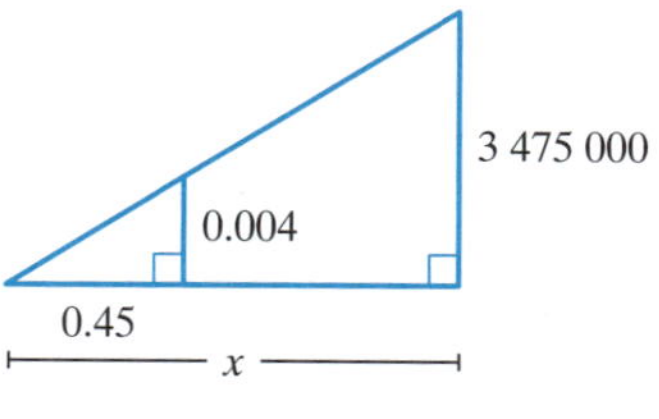

$\frac{x}{0.45} = \frac{3\,475\,000}{0.004}$

(matching sides in sim. $\triangle$s in prop.)

$x = \frac{3\,475\,000 \times 0.45}{0.004}$
$= 390\,937\,500$

$\therefore$ the distance is 390 937 500 metres.

Distance in km
$= 390\,937\,500 \div 1000$
$= 390\,937.5$

$\therefore$ Mia is 390 937.5 km from the moon.

6. $4\frac{2}{7}$ **m**

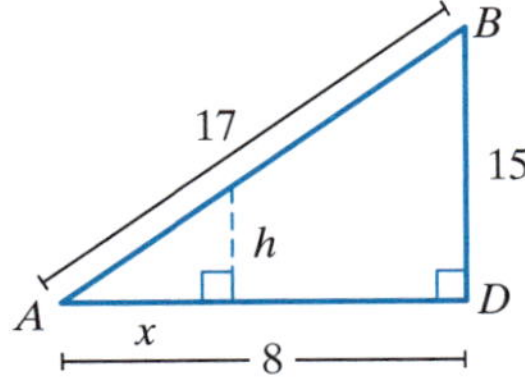

By Pythagoras,

$$BD^2 = 17^2 - 8^2$$
$$= 225$$
$$BD = 15$$
$$\frac{x}{8} = \frac{h}{15}$$

(matching sides in sim. △s in prop.)

$$x = \frac{8h}{15} \ldots ①$$

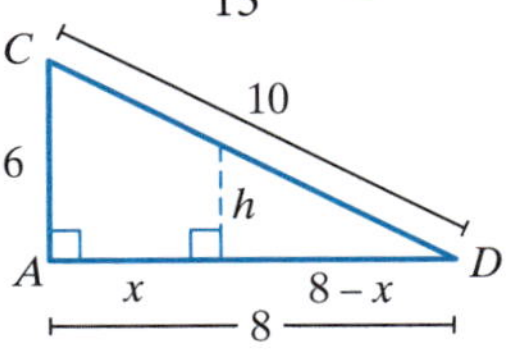

By Pythagoras,

$$CA^2 = 10^2 - 8^2$$
$$= 36$$
$$CA = 6$$
$$\frac{8-x}{8} = \frac{h}{6}$$

(matching sides in sim. △s in prop)

$$8 - x = \frac{8h}{6}$$
$$8 - x = \frac{4h}{3}$$
$$x = 8 - \frac{4h}{3} \ldots ②$$

Let ① = ②: $\frac{8h}{15} = 8 - \frac{4h}{3}$

$$8h = 120 - 20h$$
$$28h = 120$$
$$h = 4\frac{2}{7}$$

∴ the ladders cross $4\frac{2}{7}$ m from the ground.

Revision Test  11 (page 98)

1. **$5**

$$V = \pi r^2 h$$
$$= \pi \times 6^2 \times 9$$
$$= 1017.876\,02\ldots$$
$$= 1017.88 \text{ (2 dec. pl.)}$$
$$\text{Cost} = 1017.88 \times 0.0052$$
$$= 5.292\,955\,303\ldots$$
$$= 5 \text{ (nearest dollar)}$$

∴ the cost is $5.

2. **$0.15**

$$\text{Amount} = 16 \times 1.1 \times 7$$
$$= 123.2$$

∴ Max uses 123.2 litres of water.

$$\text{Cost} = 0.1232 \times 1.18$$
$$= 0.145376$$
$$= 0.15 \text{ (2 dec. pl.)}$$

∴ the cost will be $0.15.

3. **4836 L**

$$V = (2 \times 0.8 + \pi \times 0.4^2) \times 2.3$$
$$= 4.836\,106\,097\ldots$$
$$= 4.836 \text{ (3 dec. pl.)}$$

∴ the capacity is 4.836 kL, or 4836 L.

∴ the capacity is 4836 L.

4. 180π **cm³**

$$\text{CSA} = 2\pi rh$$
$$60\pi = 2 \times \pi \times r \times 5$$
$$10\pi r = 60\pi$$
$$r = 6$$

Subs. $r = 6$ and $h = 5$.

$$V = \pi r^2 h$$
$$= \pi \times 6^2 \times 5$$
$$= 180\pi$$

∴ the volume is 180π cm³.

5. **8.96 L**

Let height of bottom triangle be x.

$$x^2 = 20^2 - 16^2$$
$$= 144$$
$$x = 12$$
$$\text{Area} = \frac{1}{2} \times 32 \times 12 + 2 \times \frac{1}{2} \times 16 \times 16$$
$$= 448$$
$$\text{Vol} = 448 \times 20$$
$$= 8960$$

∴ the capacity is 8960 mL, or 8.96 L.

∴ the capacity is 8.96 L.

6. **1.17 m**

Let the height of the longer ladder be x.

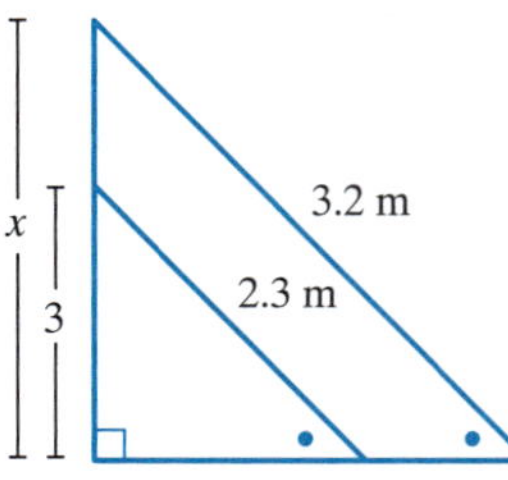

$$\frac{x}{3} = \frac{3.2}{2.3}$$
$$x = \frac{3 \times 3.2}{2.3}$$
$$= 4.173\,913\,043\ldots$$
$$= 4.17 \text{ (2 dec. pl.)}$$
$$\text{Further distance} = 4.17 - 3$$
$$= 1.17$$

∴ the ladder reaches 1.17 m further.

Revision Test  12 (page 99)

1. **1764 cm²**

Let the height of triangle be x, the hypotenuse be y and the 'depth' of the prism be z.

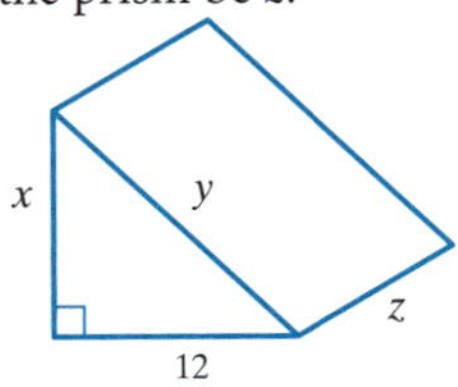

$$\text{Area of triangle} = \frac{1}{2}bh$$
$$210 = \frac{1}{2} \times 12 \times x$$
$$x = \frac{210}{6}$$
$$x = 35$$

By Pythagoras:

$$y^2 = 12^2 + 35^2$$
$$= 1369$$
$$y = 37$$
$$\text{Also, Vol.} = Ah$$
$$3360 = 210 \times z$$
$$z = \frac{3360}{210}$$
$$= 16$$
$$\text{SA} = 2 \times 210 + 12 \times 16 + 35 \times 16 + 37 \times 16$$
$$= 1764$$

∴ the surface area is 1764 cm².

2. **1357 km**

$$\text{Vol.} = 95 \times 50 \times 40$$
$$= 190\,000$$

∴ the volume is 190 000 cm³.

∴ the capacity is 190 000 mL, or 190 L.

$$\text{Distance} = 190 \div 14 \times 100$$
$$= 1357.142\,857\ldots$$
$$= 1357 \text{ (nearest whole)}$$

∴ the distance is 1357 km.

3. **0.8775 L**

Let the length be x.

Using similar triangles:

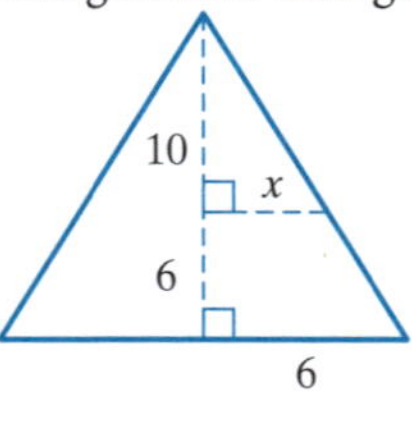

$$\frac{x}{6} = \frac{10}{16}$$
$$x = \frac{6 \times 10}{16}$$
$$= 3.75$$

∴ water forms a trapezoidal prism:

$$V = [\frac{1}{2} \times 6 \times (7.5 + 12)] \times 15$$
$$= 877.5$$

volume is 877.5 cm^3.

∴ the capacity is 877.5 mL, or 0.8775 L.

∴ the capacity is 0.8775 L.

4. $39.23

$$\text{Vol.} = \frac{1}{2} \times \pi \times 0.4^2 \times 1.4$$
$$= 0.351\,858\,377\ldots$$
$$= 0.35 \text{ (2 dec. pl.)}$$

∴ the volume is 0.35 m^3.

∴ the capacity is 0.35 kL.

$$\text{Cost} = 0.35 \times 50 \times 2.23$$
$$= 39.232\,209\,06\ldots$$
$$= 39.23 \text{ (2 dec. pl.)}$$

∴ the cost would be $39.23.

5. 214 mL

$$SA = 2\pi r^2 + 2\pi rh$$
$$66\pi = 2\pi \times 4^2 + 2\pi \times 4 \times h$$
$$66\pi = 32\pi + 8\pi h$$
$$8h = 34$$
$$h = \frac{34}{8}$$
$$= 4.25$$
$$V = \pi \times 4^2 \times 4.25$$
$$= 213.628\,300\ldots$$
$$= 214 \text{ (nearest whole)}$$

∴ the volume is 214 cm^3.

∴ the capacity is 214 mL.

6. Name the triangles, and let lengths be a and b.

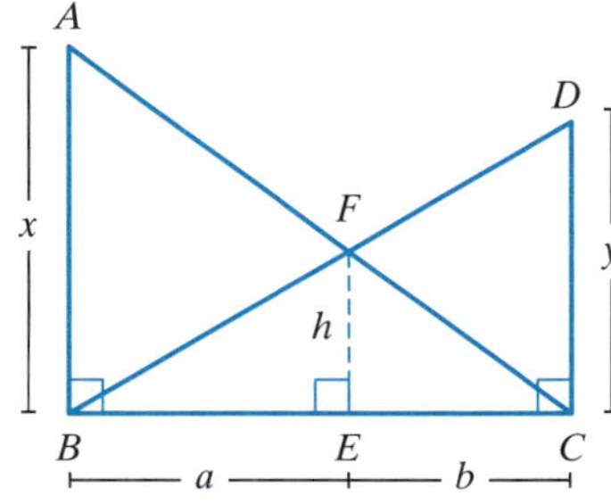

In similar triangles CFE and CAB:

$$\frac{h}{x} = \frac{b}{a+b} \ldots ①$$

In similar triangles BFE and BDC:

$$\frac{h}{y} = \frac{a}{a+b} \ldots ②$$

① + ②:

$$\frac{h}{x} + \frac{h}{y} = \frac{b}{a+b} + \frac{a}{a+b}$$
$$\frac{h}{x} + \frac{h}{y} = \frac{a+b}{a+b}$$
$$\frac{h}{x} + \frac{h}{y} = 1$$
$$\frac{1}{x} + \frac{1}{y} = \frac{1}{h}$$

Key Skill  (pages 100–101)

1. 47.8 m

Let the length of string be x.

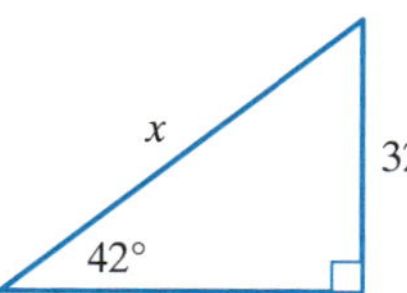

$$\frac{32}{x} = \sin 42°$$
$$x = \frac{32}{\sin 42°}$$
$$= 47.823\,2496\ldots$$
$$= 47.8 \text{ (3 sig. figs)}$$

∴ the string is 47.8 metres long.

2. 333 m

Let the distance be x.

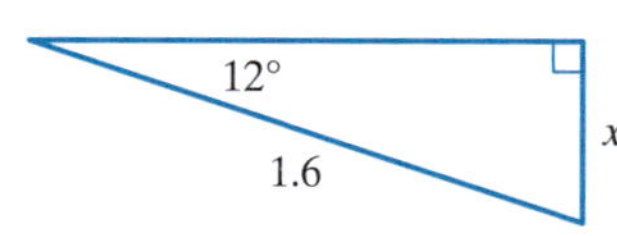

$$\frac{x}{1.6} = \sin 12°$$
$$x = 1.6 \times \sin 12°$$
$$= 0.332\,658\,705\ldots$$
$$= 0.333 \text{ (3 dec. pl.)}$$

$$\text{Distance in metres} = 0.333 \times 1000$$
$$= 333$$

∴ the end of the shaft is 333 m underground.

3. 56 m

Let the length of each wire be x.

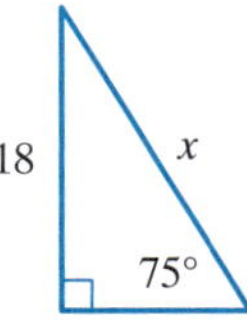

$$\frac{18}{x} = \sin 75°$$
$$x = \frac{18}{\sin 75°}$$
$$= 18.634\,971\,25\ldots$$
$$= 18.635 \text{ (3 dec. pl.)}$$

$$\text{Length of 3 wires} = 18.635 \times 3$$
$$= 55.904\,913\,74\ldots$$
$$= 56 \text{ (nearest whole)}$$

∴ 56 metres of wire is required.

4. 6.93 cm

Let the length of OM be x.

As each internal angle in a regular hexagon is 120°, then angle in triangle is 60°. As radius is 8 cm, then each side of equilateral triangle is 8 cm.

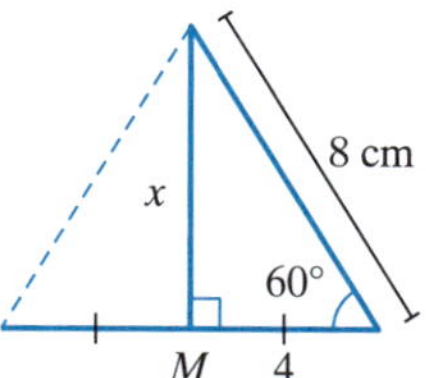

$$\frac{x}{4} = \tan 60°$$
$$x = 4 \times \tan 60°$$
$$= 6.928\,203\,23\ldots$$
$$= 6.93 \text{ (2 dec. pl.)}$$

∴ the length is 6.93 cm.

5. 8.83 km

Let the distance be x.

$$\text{Distance} = \text{Speed} \times \text{Time}$$
$$= 270 \times \frac{1}{30}$$
$$= 9$$

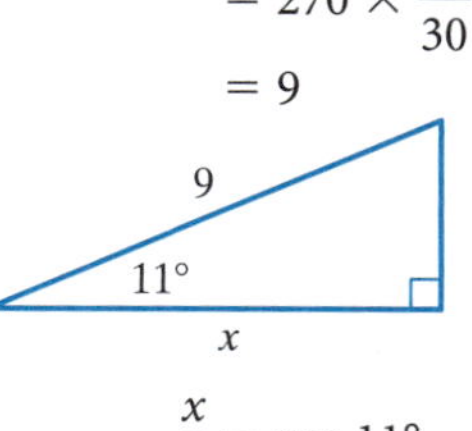

$$\frac{x}{9} = \cos 11°$$
$$x = 9 \times \cos 11°$$
$$= 8.834\,644\,651$$
$$= 8.83 \text{ (2 dec. pl.)}$$

∴ the distance is 8.83 km.

6. $46

As angle is '1 in 14', then

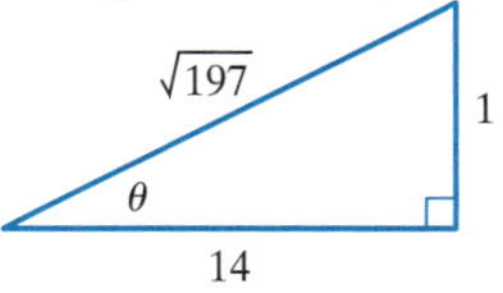

By Pythag, $\sin\theta = \frac{1}{\sqrt{197}}$

Let the length of the ramp be x.

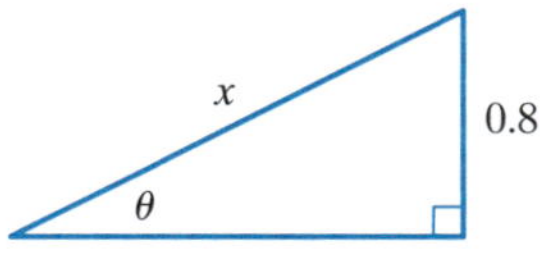

$$\frac{0.8}{x} = \sin\theta°$$
$$\frac{0.8}{x} = \frac{1}{\sqrt{197}}$$

(just like similar △s!)

$$x = 0.8 \times \sqrt{197}$$
$$= 11.228\,535\,08\ldots$$
$$= 11.229 \text{ (3 dec. pl.)}$$

$$\text{Area} = 11.229 \times 1.2$$
$$= 13.474\,242\,09\ldots$$
$$= 13.474 \text{ (3 dec pl)}$$

$$\text{Cost} = 13.474 \times 3.4$$
$$= 45.812\,423\,12\ldots$$
$$= 46 \text{ (nearest whole)}$$

∴ the cost is $46.

Key Skill  (pages 102–103)

1. 12° 38′

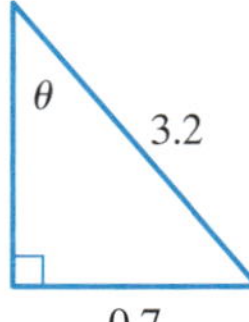

$$\sin\theta = \frac{0.7}{3.2}$$
$$\theta = 12.635\,625\,09\ldots$$
$$= 12°38'$$

∴ makes an angle of 12°38′ with the wall.

2. 84°, 48°, 48°

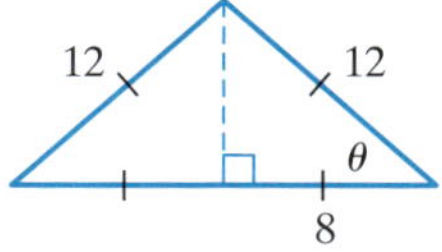

$$\cos\theta = \frac{8}{12}$$
$$\theta = 48.189\,6851\ldots$$
$$= 48 \text{ (nearest whole)}$$

As triangle is isosceles, second angle is 48°.

Third angle $= 180 - 2 \times 48$
$= 84$

∴ angles are 84°, 48° and 48°.

3. 146°19′

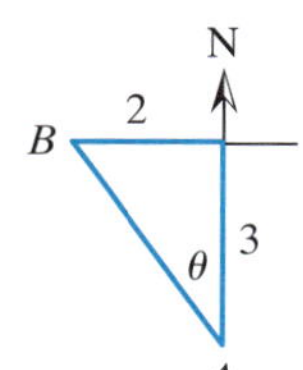

$$\tan\theta = \frac{2}{3}$$
$$\theta = 33.690\,067\,53\ldots$$
$$= 33°41' \text{ (nearest minute)}$$

Also, $180° - 33°41' = 146°19'$

∴ the boat turns 146°19′.

4. 48°35′

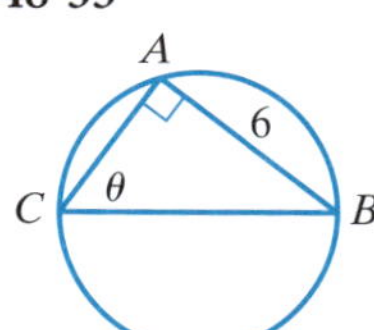

$$\text{Area} = \pi r^2$$
$$16\pi = \pi r^2$$
$$r^2 = 16$$
$$r = 4 \quad (r > 0)$$

$\therefore CB = 8$

$$\sin\theta = \frac{6}{8}$$
$$\theta = 48.590\,377\,89\ldots$$
$$= 48°35' \text{ (nearest minute)}$$

∴ the angle is 48°35′.

5. 8°56′

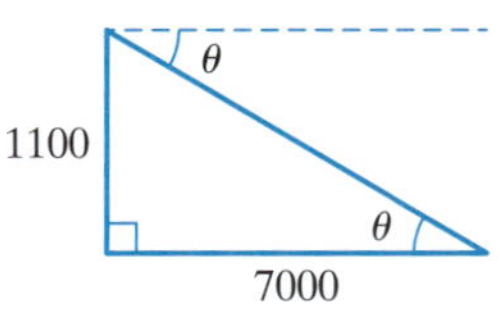

$$\tan\theta = \frac{1100}{7000}$$
$$\theta = 8.930\,5901\ldots$$
$$= 8°56' \text{ (nearest minute)}$$

∴ the angle of approach is 8°56′.

6. 19°26′

Let AC be x.

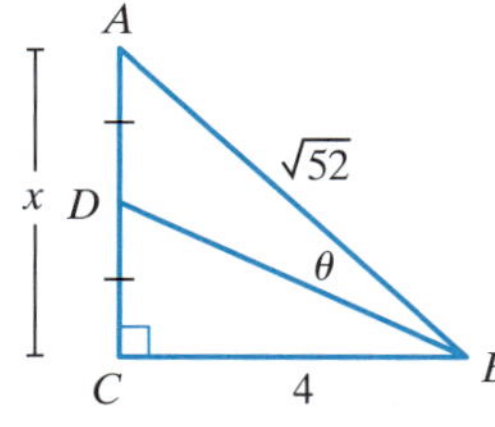

By Pythagoras:

$$x^2 = (\sqrt{52})^2 - 4^2$$
$$= 52 - 16$$
$$= 36$$
$$x = 6$$

$\therefore AD = DC = 3$

$$\tan\angle CBD = \frac{3}{4}$$
$$\angle CBD = 36.869\,897\,65\ldots$$
$$\tan\angle CBA = \frac{6}{4}$$
$$\angle CBD = 56.309\,932\,47\ldots$$
$$\therefore \theta = 56.309\,932\,47 - 36.869\,897\,65$$
$$= 19.440\,034\,83$$
$$= 19°26' \text{ (nearest minute)}$$

∴ the angle is 19°26′.

Key Skill  (pages 104–105)

1. 33.61 m

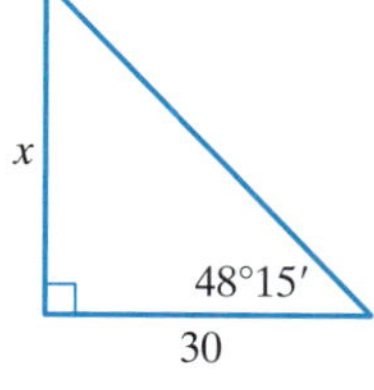

$$\frac{x}{30} = \tan 48°15'$$
$$= 30 \times \tan 48°15'$$
$$= 33.612\,160\,23\ldots$$
$$= 33.61 \text{ (2 dec. pl.)}$$

∴ the tree is 33.61 m tall.

2. 48°1′

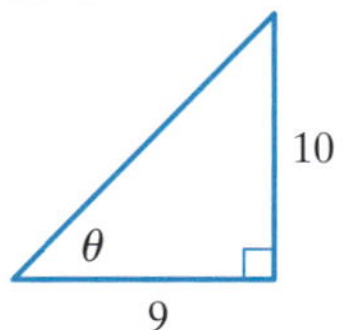

$$\tan\theta = \frac{10}{9}$$
$$\theta = 48.012\,7875\ldots$$
$$= 48°1'$$

∴ the angle of elevation is 48°1′.

3. 52 m

Let the difference in the two heights be x.

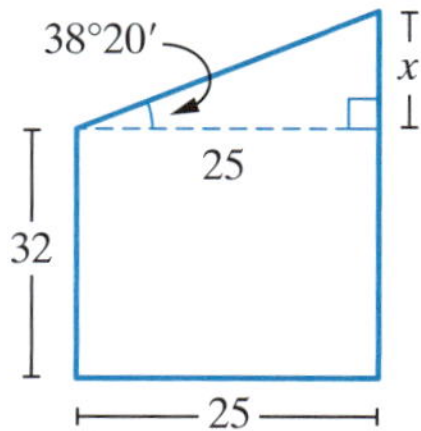

$$\frac{x}{25} = \tan 38°20'$$
$$= 25 \times \tan 38°20'$$
$$= 19.767\,437\,06\ldots$$
$$= 20 \text{ (nearest whole)}$$

Height of taller building $= 32 + 20$
$= 52$

∴ the building is 52 metres tall.

4. 7523 m

Let the different distances be x and y.

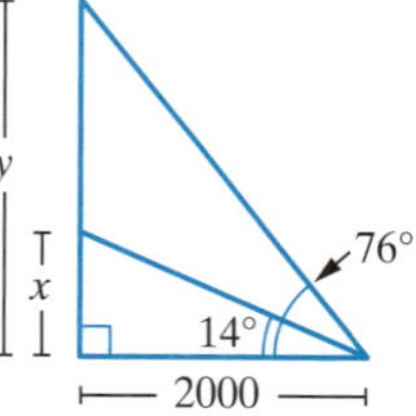

$$\frac{x}{2000} = \tan 14°$$
$$x = 2000 \times \tan 14°$$
$$= 498.656\,005\ldots$$
$$= 498.66 \text{ (2 dec. pl.)}$$
$$\frac{y}{2000} = \tan 76°$$
$$y = 2000 \times \tan 76°$$
$$= 8021.561\,867\ldots$$
$$= 8021.56 \text{ (2 dec pl)}$$

Difference $= 8021.56 - 498.66$
$= 7522.9$
$= 7523$ (nearest whole)

∴ the rocket travelled 7523 metres.

5. **31°**

Let the height of the building be x, and let the angle of elevation be θ:

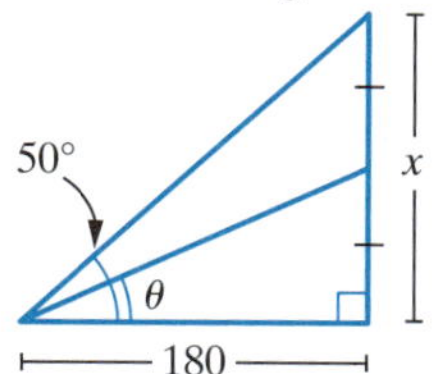

$$\frac{x}{180} = \tan 50°$$
$$x = 180 \times \tan 50°$$
$$= 214.515\,6467\ldots$$
$$= 214.52 \text{ (2 dec. pl.)}$$
$$\text{Height halfway} = 214.52 \div 2$$
$$= 107.257\,8233\ldots$$
$$= 107.26 \text{ (2 dec. pl.)}$$
$$\tan\theta = \frac{107.26}{180}$$
$$\theta = 30.789\,733\,03\ldots$$
$$= 31° \text{ (nearest degree)}$$

$\therefore$ the angle of elevation would be 31°.

6. **1287 km/h**

Let the distance travelled be x.

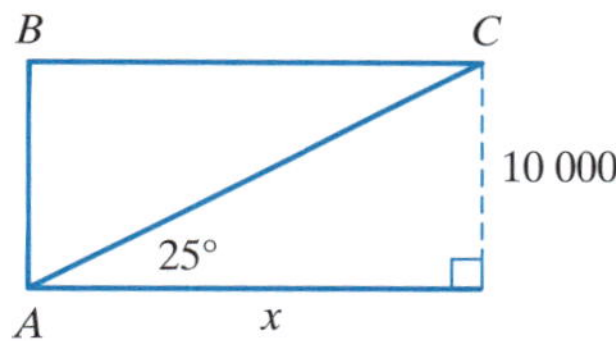

$$\frac{10\,000}{x} = \tan 25°$$
$$x = \frac{10\,000}{\tan 25°}$$
$$= 21\,445.069\,21\ldots$$
$$= 21\,445.07 \text{ (2 dec. pl.)}$$
$$\text{Distance in km} = 21\,445.07 \div 1000$$
$$= 21.445\,07$$
$$\text{Speed} = \text{Distance} \div \text{Time}$$
$$= 21.445\,07 \div \frac{1}{60}$$
$$= 1286.704\,152\ldots$$
$$= 1287 \text{ (nearest whole)}$$

$\therefore$ the speed is 1287 km/h.

Key Skill  43 (pages 106–107)

1. **61°**

Let the angle of depression be θ.

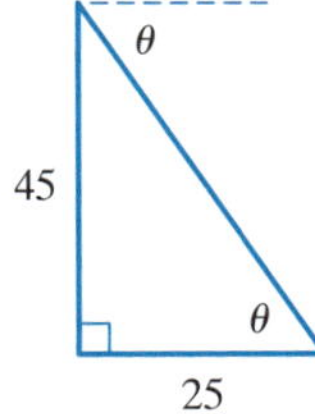

$$\tan\theta = \frac{45}{25}$$
$$\theta = 60.945\,3959\ldots$$
$$= 61 \text{ (nearest whole)}$$

$\therefore$ the angle of depression is 61°.

2. **6209 m**

Let the distance be x.

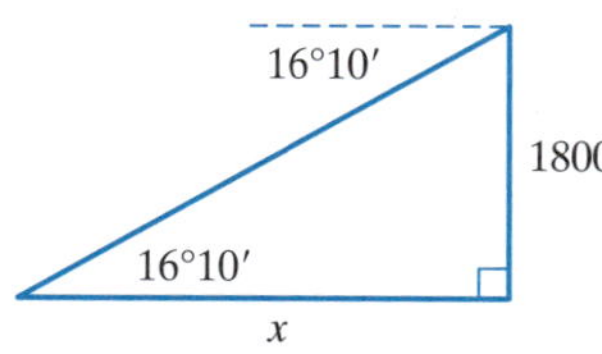

$$\frac{1800}{x} = \tan 16°10'$$
$$x = \frac{1800}{\tan 16°10'}$$
$$= 6209.121\,526\ldots$$
$$= 6209 \text{ (nearest whole)}$$

$\therefore$ the distance is 6209 m.

3. **38°40′**

Let the angle of depression be θ.

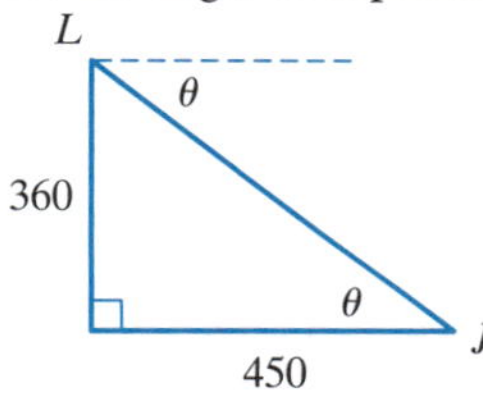

$$\tan\theta = \frac{360}{450}$$
$$\theta = 38.659\,808\,25\ldots$$
$$= 38°40' \text{ (nearest minute)}$$

$\therefore$ the angle of depression is 38°40′.

4. **2°39′**

Let the angle of depression be θ.

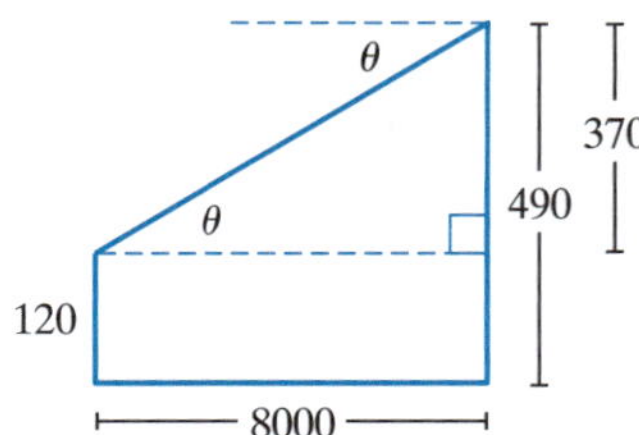

$$\tan\theta = \frac{370}{8000}$$
$$= 2.648\,042\,769\ldots$$
$$= 2°39' \text{ (nearest minute)}$$

$\therefore$ the angle of depression is 2°39′.

5. **48°39′**

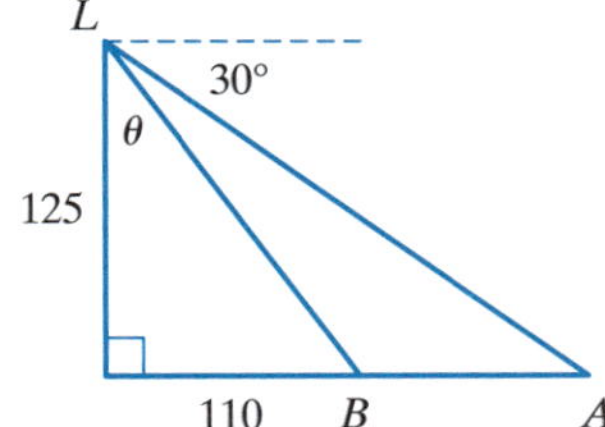

$$\tan\theta = \frac{110}{125}$$
$$\theta = 41.347\,777\,22\ldots$$
$$= 41°21' \text{ (nearest minute)}$$

$\therefore$ angle of depression of
$$B = 90° - 41°21'$$
$$= 48°39'$$

$\therefore$ the angle of depression is 48°39′

6. **67 m**

Let distances from the hill be x and y.

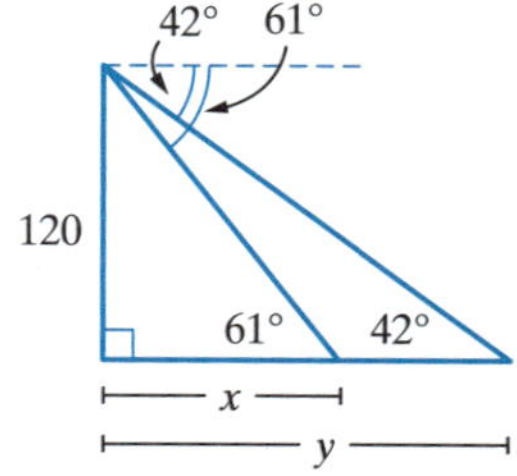

$$\frac{120}{x} = \tan 61°$$
$$x = \frac{120}{\tan 61°}$$
$$\frac{120}{y} = \tan 42°$$
$$y = \frac{120}{\tan 42°}$$
$$\text{Difference} = \frac{120}{\tan 42°} - \frac{120}{\tan 61°}$$
$$= 66.756\,415\,61\ldots$$
$$= 67 \text{ (nearest whole)}$$

$\therefore$ the width is 67 metres.

Key Skill  44 (pages 108–109)

1. **179 km**

Let the distance be x.

$348 - 270 = 78$.

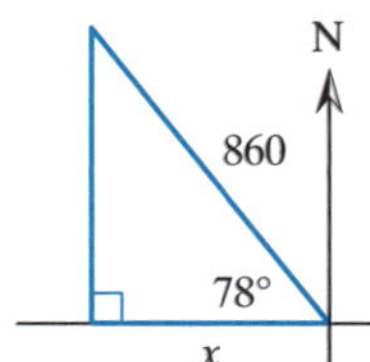

$$\frac{x}{860} = \cos 78°$$
$$x = 860 \times \cos 78°$$
$$= 178.804\,0541\ldots$$
$$= 179 \text{ (nearest whole)}$$

$\therefore$ the ship is 179 km west of Fremantle.

2. **207°**

Let the angle be θ.

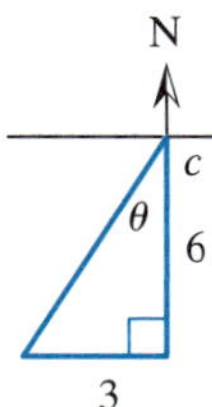

$$\tan\theta = \frac{3}{6}$$
$$\theta = 26.565\,051\,18\ldots$$
$$= 27 \text{ (nearest whole)}$$

$180 + 27 = 207$

$\therefore$ the bearing is 207°.

3. **96.35 km**

Let the distance be x.

$72 \times 2 = 144$.

$228 - 180 = 48$.

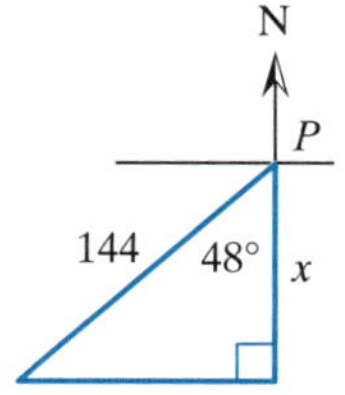

$$\frac{x}{144} = \cos 48°$$
$$x = 144 \times \cos 48°$$
$$= 96.354\,807\,32\ldots$$
$$= 96.35 \text{ (2 dec. pl.)}$$

$\therefore$ Jans is 96.35 km south.

4. **221°**

Let the angle be θ.

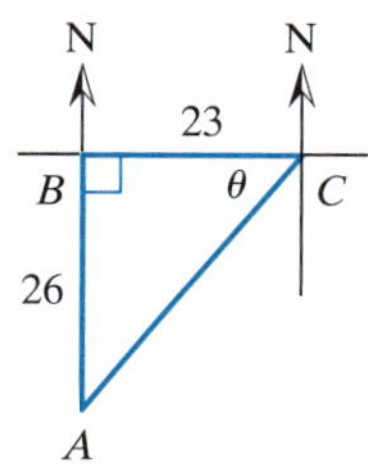

$$\tan\theta = \frac{26}{23}$$
$$\theta = 48.503\,531\,64\ldots$$
$$= 49 \text{ (nearest whole)}$$

$270 - 49 = 221$

$\therefore$ the bearing is 221°.

5. **13 m, 023°**

Let the distance be x, the angle be θ.

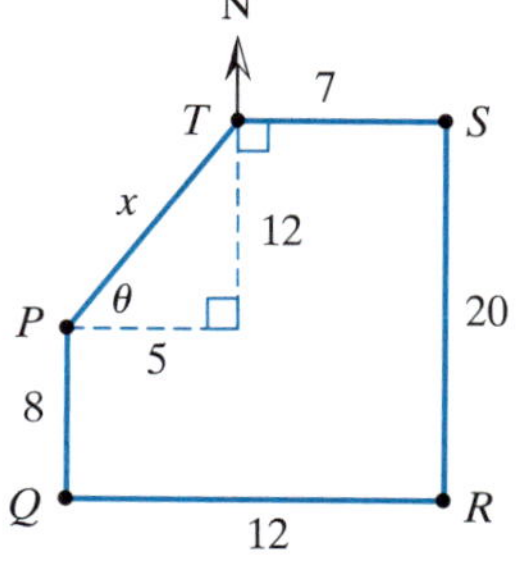

By Pythagoras,
$$x^2 = 5^2 + 12^2$$
$$= 169$$
$$= 13$$
$$\tan\theta = \frac{12}{5}$$
$$\theta = 67.380\,135\,05\ldots$$
$$= 67 \text{ (nearest whole)}$$

The required angle $= 90° - 67°$
$= 23°$

$\therefore$ the distance is 13 metres, the bearing is 023°.

6. **15 km/h**

Let the distance be x.

10 am minus 7:30 am is 2.5 h.

$$\text{Distance} = 80 \times 2.5$$
$$= 200$$

$270 - 230 = 40$.

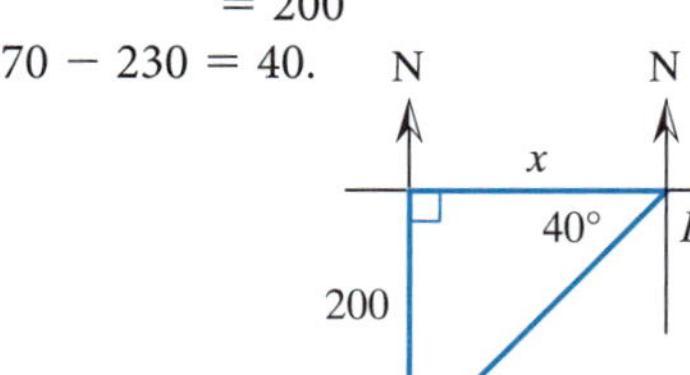

$$\frac{200}{x} = \tan 40°$$
$$x = \frac{200}{\tan 40°}$$
$$= 238.350\,7185\ldots$$
$$= 238.35 \text{ (2 dec. pl.)}$$
$$\text{Speed} = 238.35 \div 2.5$$
$$= 95.340\,287\,41\ldots$$
$$= 95 \text{ (nearest whole)}$$
$$\text{Difference} = 95 - 80$$
$$= 15$$

$\therefore$ B is travelling 15 km/h faster than A.

Key Skill 45 (pages 110–111)

1. **6.403 km**

Let the distance from start to Lee be x.

$$\text{Walk distance} = 6 \times \frac{2}{3}$$
$$= 4$$
$$\text{Jog distance} = 10 \times \frac{1}{2}$$
$$= 5$$

$120 - 90 = 30$ and $270 - 210 = 60$.

$\therefore 30 + 60 = 90$.

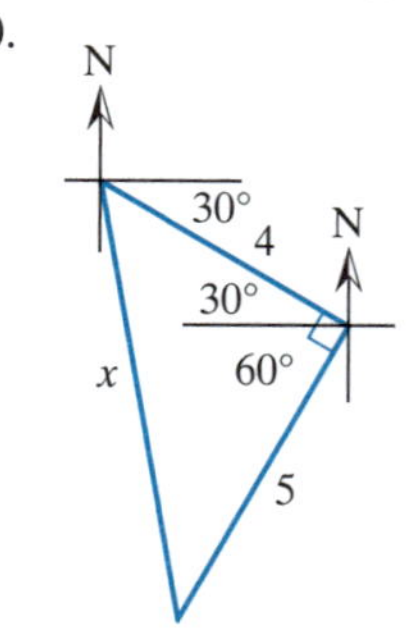

By Pythagoras:
$$x^2 = 4^2 + 5^2$$
$$= 41$$
$$x = \sqrt{41}$$
$$= 6.403\,124\,237\ldots$$
$$= 6.403 \text{ (3 dec. pl.)}$$

$\therefore$ Lee is 6.403 km from the start position.

2. **164°**

Let the angle be θ.

First distance $= 12 \times 2 = 24$

Second distance $= 16$

$310 - 270 = 40$ and $90 - 40 = 50$.

$\therefore 40 + 50 = 90$.

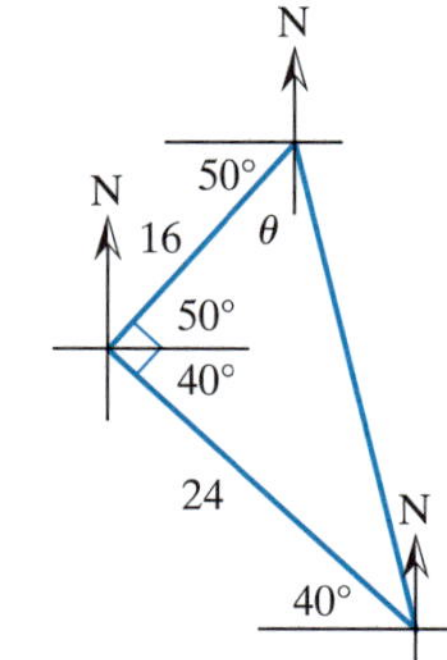

$$\tan\theta = \frac{24}{16}$$
$$\theta = 56.309\,932\,47\ldots$$
$$= 56 \text{ (nearest whole)}$$

$270 - (50 + 56) = 164$

$\therefore$ the bearing is 164°.

3. **984 km**

Let the distance from start be x.

$$1^{st} \text{ leg} = 360 \times 2\frac{1}{6}$$
$$= 780$$
$$2^{nd} \text{ leg} = 360 \times 1\frac{2}{3}$$
$$= 600$$

$137 - 90 = 47$ and $270 - 227 = 43$.

$\therefore 47 + 43 = 90$.

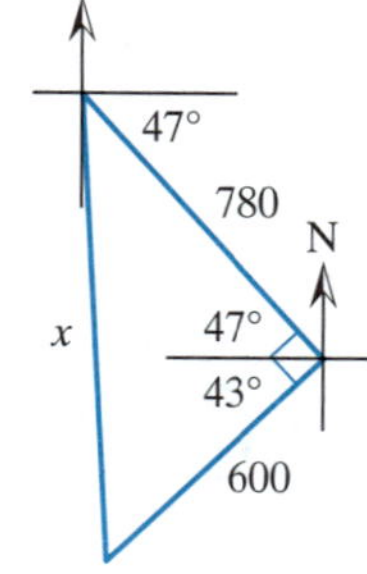

By Pythagoras:
$$x^2 = 780^2 + 600^2$$
$$= 968\,400$$
$$x = \sqrt{968\,400}$$
$$= 984.073\,168\ldots$$
$$= 984 \text{ (nearest whole)}$$

$\therefore$ the plane is 984 km from its starting point.

4. **080°**
Let the angle be θ.
$180 - 125 = 55$.
$\therefore 55 + 35 = 90$.

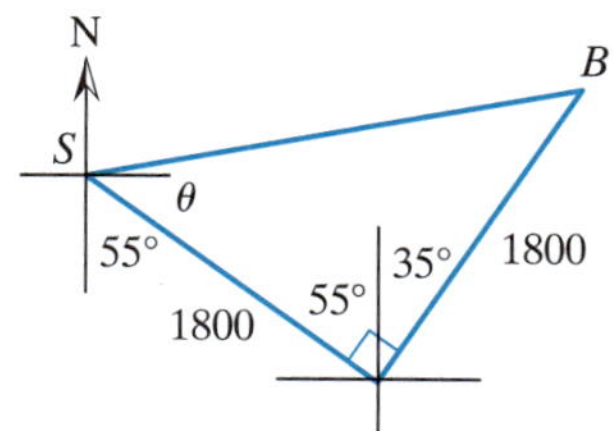

As △ is isosceles and right-angled, then $\theta = 45°$.
$180 - (45 + 55) = 80$
∴ the bearing is 080°.

5. **38 km**
Let the distance be x.
$125 - 90 = 35$ and $270 - 215 = 55$.
$\therefore 35 + 55 = 90$.
$90 - (28 + 55) = 7$.

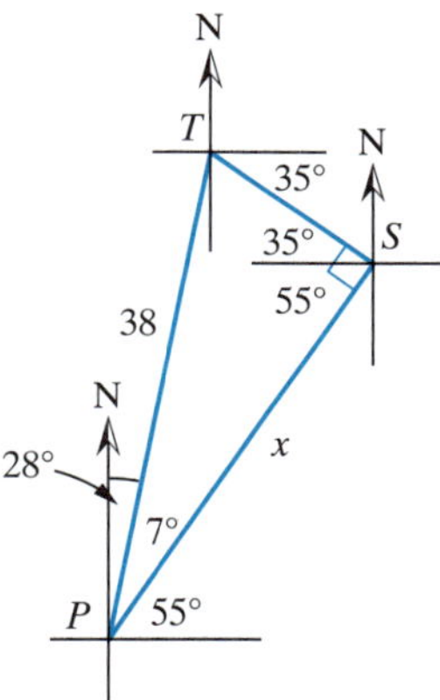

$$\frac{x}{38} = \cos 7°$$
$$x = 38 \times \cos 7°$$
$$= 37.716\,753\,76\ldots$$
$$= 38 \text{ (nearest whole)}$$
∴ the distance is 38 km.

6. **1:56 pm**
Let the distance be x.
$180 - 110 = 70$ and
$180 - (50 + 70) = 60$.
Also, $70 + 20 = 90$.

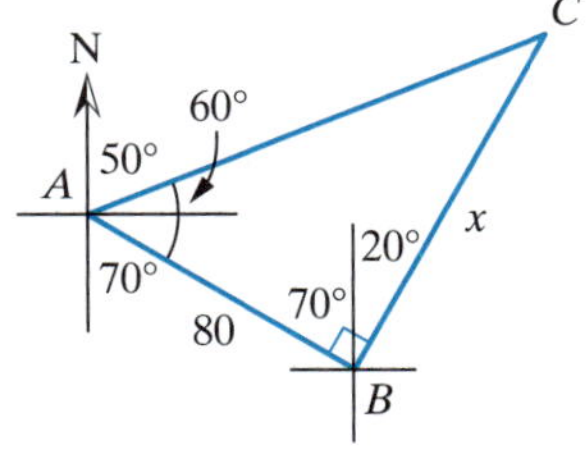

$$\frac{x}{80} = \tan 60°$$
$$x = 80 \times \tan 60°$$
$$= 138.564\,0646\ldots$$
$$= 138.56 \text{ (2 dec. pl.)}$$
$$\text{Total distance} = 80 + 138.56$$
$$= 218.56$$
$$\text{Time} = 218.56 \div 90$$
$$= 2.428\,489\,607\ldots$$
$$= 2 \text{ h } 26 \text{ min (nearest minute)}$$
Arrival time = 11:30 plus 2 h 26 min
= 1:56 pm
∴ Regan arrived at 1:56 pm.

Key Skill  46 (pages 112–113)

1. **34 m**
Let the two distances be x and y.

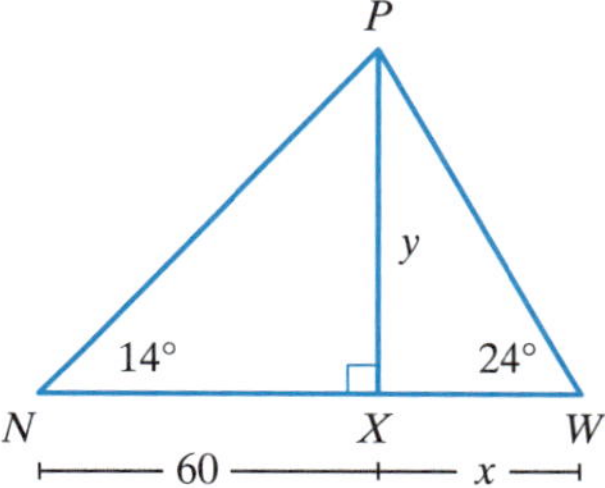

$$\frac{y}{60} = \tan 14°$$
$$y = 60 \times \tan 14°$$
Also, $\frac{y}{x} = \tan 24°$
$$x = \frac{y}{\tan 24°}$$
$$= \frac{60 \times \tan 14°}{\tan 24°}$$
$$= 33.599\,991\,79\ldots$$
$$= 34 \text{ (nearest metre)}$$
∴ William is 34 metres from the pole.

2. **223 km**
Let the two distances be x and y.
$345 - 270 = 75$, $90 - 80 = 10$.

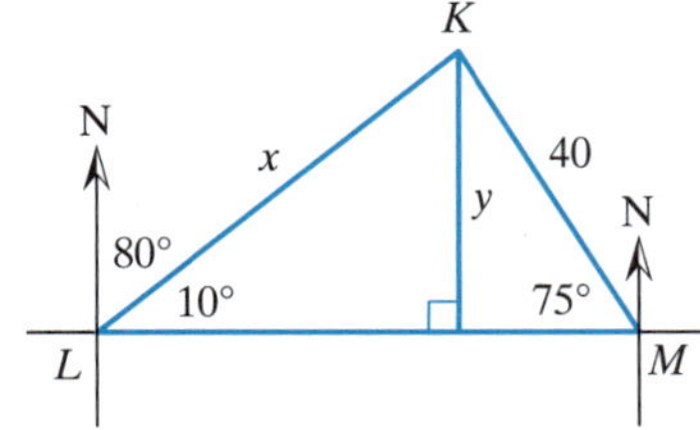

$$\frac{y}{40} = \sin 75°$$
$$y = 40 \times \sin 75°$$
Also, $\frac{y}{x} = \sin 10°$
$$x = \frac{y}{\sin 10°}$$
$$= \frac{40 \times \sin 75°}{\sin 10°}$$
$$= 222.501\,8055\ldots$$
$$= 223 \text{ (nearest metre)}$$
∴ the distance is 223 km.

3. **46 m**
Let the two distances be x and y.

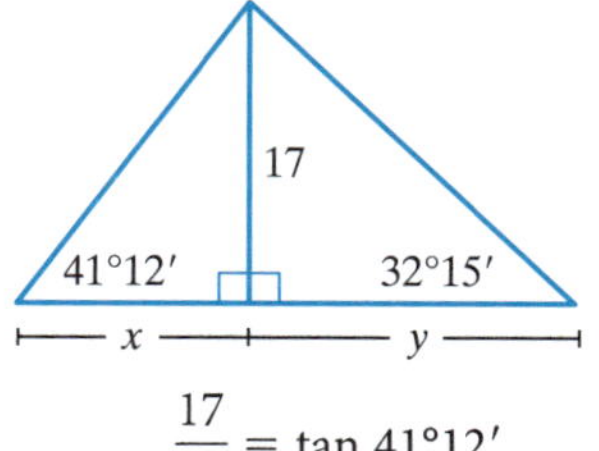

$$\frac{17}{x} = \tan 41°12'$$
$$x = \frac{17}{\tan 41°12'}$$
Also, $\frac{17}{y} = \tan 32°15'$
$$y = \frac{17}{\tan 32°15'}$$
Total distance
$$= \frac{17}{\tan 41°12'} + \frac{17}{\tan 32°15'}$$
$$= 46.362\,312\,87\ldots$$
$$= 46 \text{ (nearest whole)}$$
∴ the girls are 46 metres apart.

4. **41 km**
Let the two distances be x and y.
$180 - 130 = 50$.

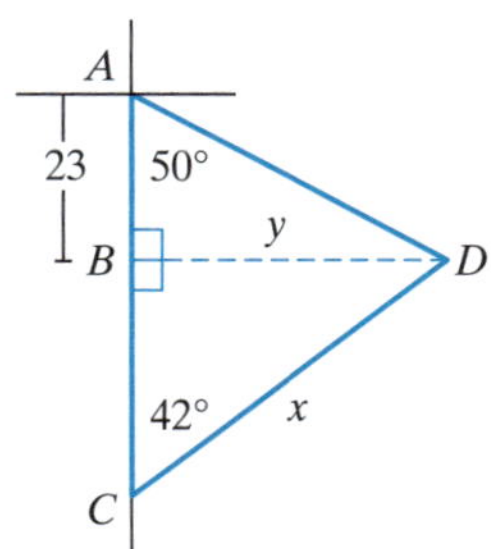

$$\frac{y}{23} = \tan 50°$$
$$y = 23 \times \tan 50°$$
Also, $\frac{y}{x} = \sin 42°$
$$x = \frac{y}{\sin 42°}$$
$$= \frac{23 \times \tan 50°}{\sin 42°}$$
$$= 40.964\,099\,34\ldots$$
$$= 41 \text{ (nearest whole)}$$
∴ the distance is 41 km.

5. 37 m, 59 m

Let the two heights be x and y.

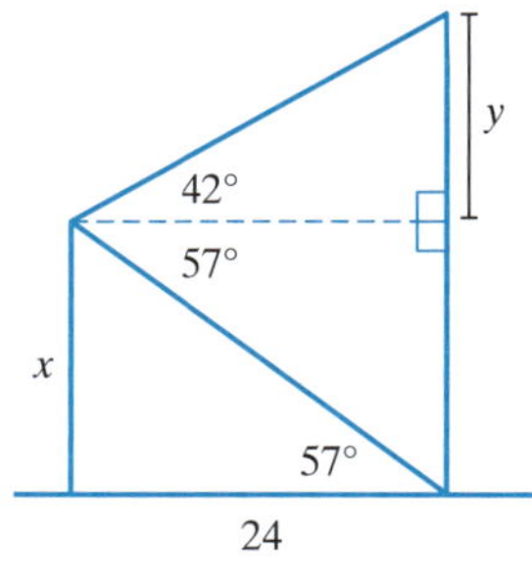

$$\frac{x}{24} = \tan 57°$$
$$x = 24 \times \tan 57°$$
$$= 36.956\,759\,13\ldots$$
$$= 37 \text{ (nearest whole)}$$

Also, $\frac{y}{24} = \tan 42°$
$$y = 24 \times \tan 42°$$

Height of tall building $= x + y$
$$= 58.566\,456\,19\ldots$$
$$= 59 \text{ (nearest whole)}$$

∴ the heights are 37 m and 59 m.

6. 144 m

Let the height be h and two lengths be x and y.

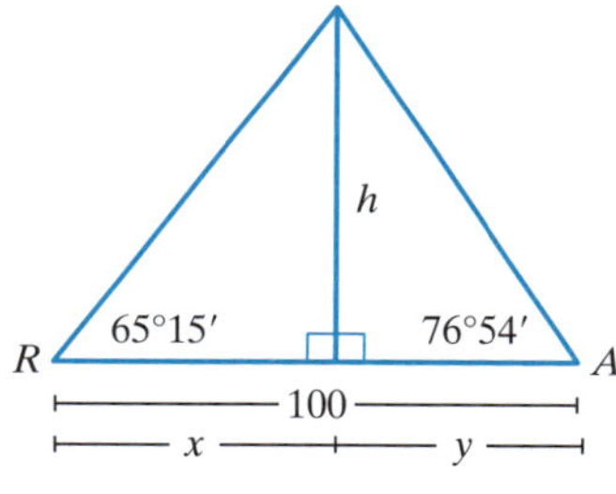

$$\frac{h}{x} = \tan 65°15'$$
$$x = \frac{h}{\tan 65°15'}$$

Also, $\frac{h}{y} = \tan 76°54'$
$$y = \frac{h}{\tan 76°54'}$$

Now, $x + y = 100$:
$$\frac{h}{\tan 65°15'} + \frac{h}{\tan 76°54'} = 100$$
$$h(\tan 76°54' + \tan 65°15')$$
$$= 100 \times \tan 65°15' \times \tan 76°54'$$
$$h = \frac{100 \times \tan 76°54' \times \tan 65°15'}{\tan 65°15' + \tan 76°54'}$$
$$= 144.151\,7067\ldots$$
$$= 144 \text{ (nearest whole)}$$

∴ the height is 144 m.

Key Skill  47 (pages 114–115)

1. 28 m

Let the two distances be x and y.

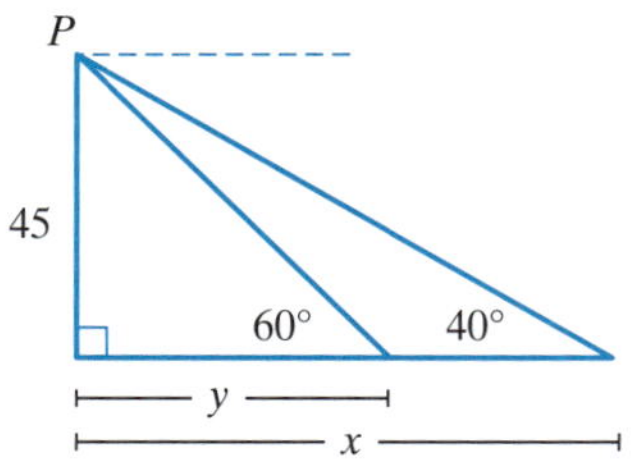

$$\frac{45}{y} = \tan 60°$$
$$y = \frac{45}{\tan 60°}$$

Also, $\frac{45}{x} = \tan 40°$
$$x = \frac{45}{\tan 40°}$$

Distance walked
$$= \frac{45}{\tan 40°} - \frac{45}{\tan 60°}$$
$$= 27.648\,149\,55\ldots$$
$$= 28 \text{ (nearest metre)}$$

∴ Tony walked 28 metres.

2. 292 km

Let the distances be x and y.

$280 - 270 = 10$, $340 - 270 = 70$.

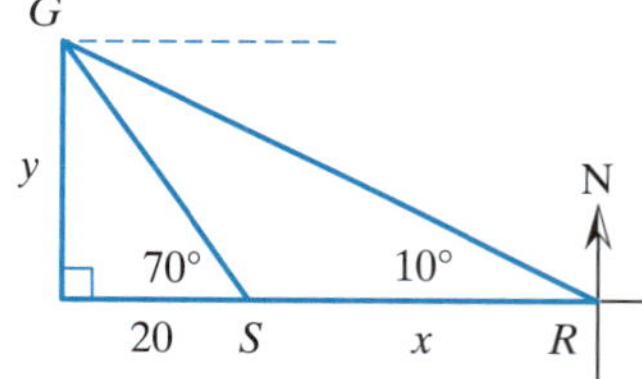

$$\frac{y}{20} = \tan 70°$$
$$y = 20 \times \tan 70°$$

Also, $\frac{y}{x + 20} = \tan 10°$
$$x + 20 = \frac{y}{\tan 10°}$$
$$x = \frac{20 \times \tan 70°}{\tan 10°} - 20$$
$$= 291.634\,3748\ldots$$
$$= 292 \text{ (nearest whole)}$$

∴ rest area to service station was 292 km.

3. 10 m

Let the height of the flagpole be x and the height of the building be y.

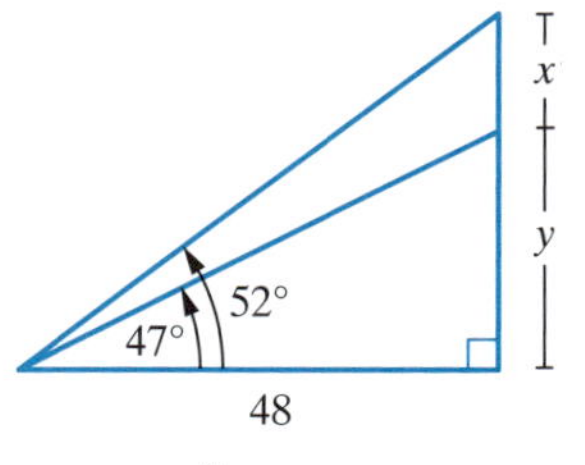

$$\frac{y}{48} = \tan 47°$$
$$y = 48 \times \tan 47° \ldots ①$$

Also, $\frac{x + y}{48} = \tan 52°$
$$x + y = 48 \times \tan 52° \ldots ②$$

Subs. ① into ②:
$$x + 48 \times \tan 47° = 48 \times \tan 52°$$
$$x = 48 \times \tan 52° - 48 \times \tan 47°$$
$$= 48(\tan 52° - \tan 47°)$$
$$= 9.963\,500\,264\ldots$$
$$= 10 \text{ (nearest whole)}$$

∴ the flagpole is 10 m high.

4. 6 m

Let the two distances be x and y.

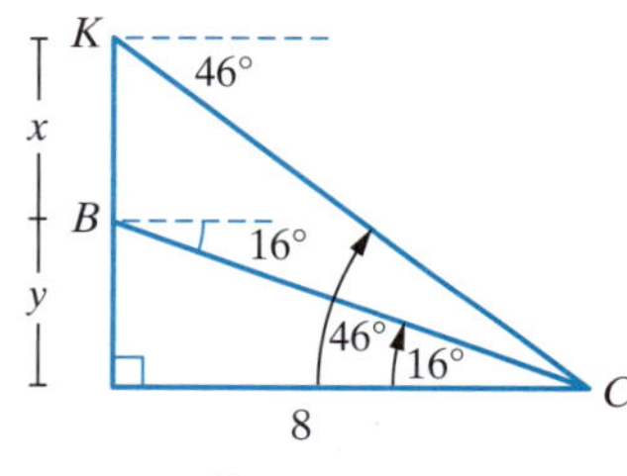

$$\frac{y}{8} = \tan 16°$$
$$y = 8 \times \tan 16° \ldots ①$$

Also, $\frac{x + y}{8} = \tan 46°$
$$x + y = 8 \times \tan 46° \ldots ②$$

Subs. ① into ②:
$$x + 8 \times \tan 16° = 8 \times \tan 46°$$
$$x = 8 \times \tan 46° - 8 \times \tan 16°$$
$$= 8(\tan 46° - \tan 16°)$$
$$= 5.990\,279\,424\ldots$$
$$= 6 \text{ (nearest whole)}$$

∴ Ken's seat is 6 metres above Ben's seat.

5. 47 m

Let the height be x and the distance be y.

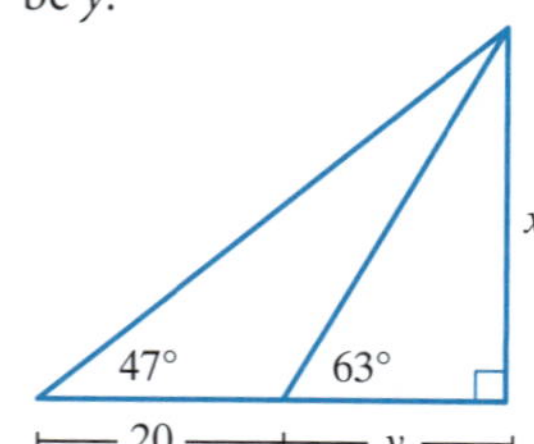

$\frac{x}{y} = \tan 63°$

$x = y \times \tan 63° \ldots$ ①

Also, $\frac{x}{y + 20} = \tan 47°$

$x = (y + 20) \times \tan 47°$

$= y \times \tan 47° + 20 \times \tan 47° \ldots$ ②

① = ②:

$y \times \tan 63°$

$= y \times \tan 47° + 20 \times \tan 47°$

$y(\tan 63° - \tan 47°) = 20 \times \tan 47°$

$y = \frac{20 \times \tan 47}{\tan 63° - \tan 47°}$

Subs. in ①:

$x = \frac{20 \times \tan 47°}{\tan 63° - \tan 47°} \times \tan 63°$

$= 47.282\,482\,23\ldots$

$= 47$ (nearest whole)

∴ the tree is 47 metres high.

6. 9:57 am

Let the distances be x and y.

$180 - 170 = 10$ and

$180 - 140 = 40$.

Distance $= 80 \times \frac{3}{4}$

$= 60$

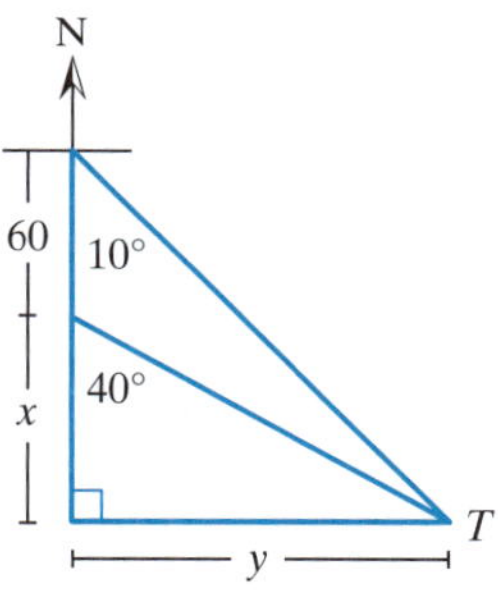

$\frac{y}{x} = \tan 40°$

$y = x \times \tan 40° \ldots$ ①

Also, $\frac{y}{x + 60} = \tan 10°$

$y = (x + 60) \times \tan 10° \ldots$ ②

① = ②:

$x \times \tan 40° = (x + 60) \times \tan 10°$

$x \times \tan 40°$

$= x \tan 10° + 60 \times \tan 10°$

$x \times \tan 40° - x \tan 10°$

$= 60 \times \tan 10°$

$x(\tan 40° - \tan 10°) = 60 \times \tan 10°$

$x = \frac{60 \times \tan 10°}{\tan 40° - \tan 10°}$

$= 15.962\,666\,59\ldots$

$= 15.96$ (2 dec. pl.)

Time $= 15.96 \div 80$

$= 0.199\,533\,332\ldots$

$= 12$ min (nearest minute)

9:45 am plus 12 min = 9:57 am.

∴ Wade will be west of the tower at 9:57 am.

Revision Test 13 (page 116)

1. 1.04 m

Let the distance from the wall be x.

$\frac{x}{2.8} = \cos 68°6'$

$x = 2.8 \times \cos 68°6'$

$= 1.044\,365\,791\ldots$

$= 1.04$ (2 dec pl)

∴ the ladder is 1.04 m from the wall.

2. 51°, 51°, 78°

The triangle is isosceles.

Let an angle be θ.

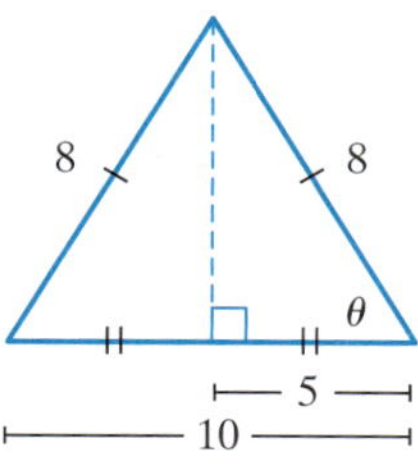

$\cos \theta = \frac{5}{8}$

$\theta = 51.317\,812\,55\ldots$

$= 51$

∴ the base angles are 51°.

Also, $180 - 2 \times 51 = 78$.

∴ the other angle is 78°.

∴ the angles are 51°, 51° and 78°.

3. 2.77 m

Let the lengths be x and y.

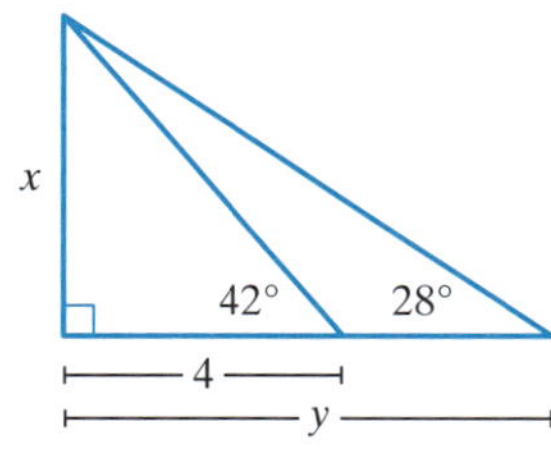

$\frac{x}{4} = \tan 42°$

$x = 4 \times \tan 42°$

Also, $\frac{x}{y} = \tan 28°$

$x = y \times \tan 28°$

∴ $y \times \tan 28° = 4 \times \tan 42°$

$y = \frac{4 \times \tan 42°}{\tan 28°}$

$= 6.773\,654\,862\ldots$

$= 6.77$ (3 sig. figs)

Difference $= 6.77 - 4$

$= 2.77$

∴ the shadow is 2.77 m longer.

4. 27 m

Let the difference in heights be x.

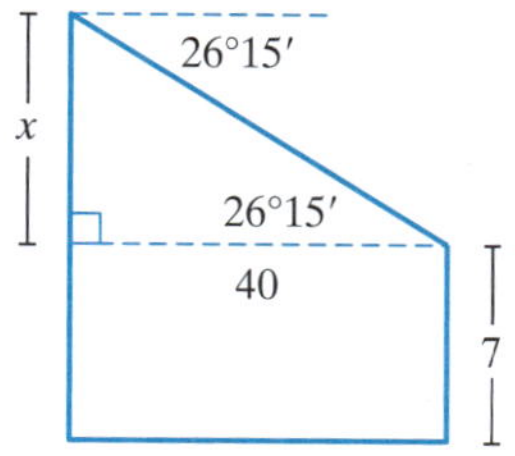

$\frac{x}{40} = \tan 26°15'$

$x = 40 \times \tan 26°15'$

$= 19.725\,817\,04\ldots$

$= 20$ (nearest whole)

Height $= 20 + 7$

$= 27$

∴ the taller pole is 27 metres high.

5. 3.54 m

Let the distances be x and y.

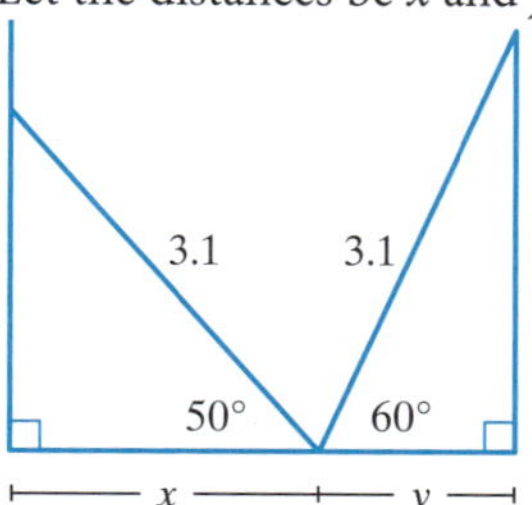

$\frac{x}{3.1} = \cos 50°$

$x = 3.1 \times \cos 50°$

$\frac{y}{3.1} = \cos 60°$

$y = 3.1 \times \cos 60°$

Width of alley

$= 3.1 \times \cos 50° + 3.1 \times \cos 60°$

$= 3.542\,641\,59\ldots$

$= 3.54$ (2 dec. pl.)

∴ the alley is 3.54 metres wide.

6. 11 m

Let the heights be x and y.

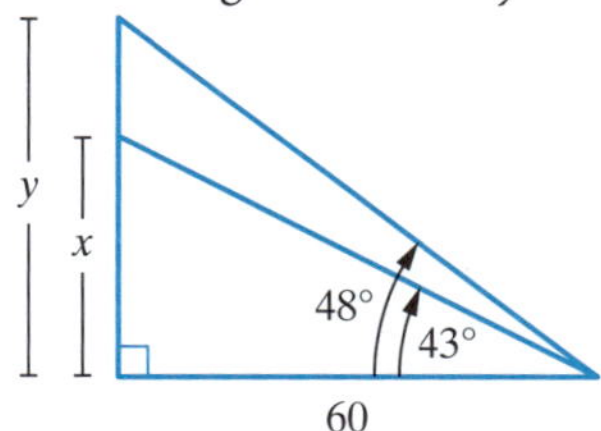

$\frac{y}{60} = \tan 48°$

$y = 60 \times \tan 48°$

$\frac{x}{60} = \tan 43°$

$x = 60 \times \tan 43°$

Height of sign

$= 60 \times \tan 48° - 60 \times \tan 43°$

$= 10.685\,845\,72\ldots$

$= 11$ (nearest whole)

∴ the sign is 11 metres high.

Revision Test  (page 117)

1. 006°

Let the angle be θ.

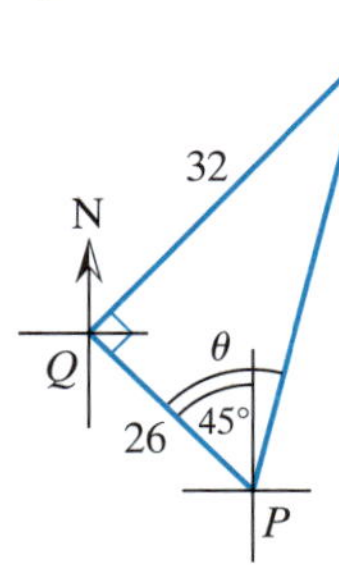

$$\tan\theta = \frac{32}{26}$$
$$\theta = 50.906\,141\,11\ldots$$
$$= 51 \text{ (nearest whole)}$$
$$\text{Bearing} = 51 - 45$$
$$= 6$$

$\therefore$ the bearing is 006°.

2. 175 km

Let the distance be x.

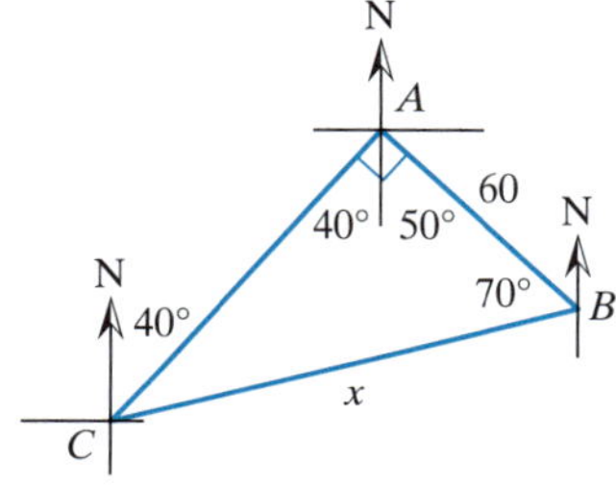

$$\frac{60}{x} = \cos 70°$$
$$x = \frac{60}{\cos 70°}$$
$$= 175.428\,264\ldots$$
$$= 175 \text{ (nearest whole)}$$

$\therefore$ the distance is 175 km.

3. 21 m

Let the lengths be x, y and z.

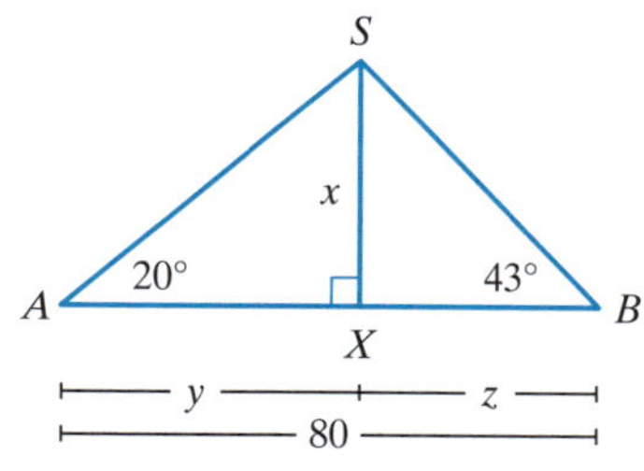

$$\frac{x}{y} = \tan 20°$$
$$y = \frac{x}{\tan 20°}$$
$$\text{Also, } \frac{x}{z} = \tan 43°$$
$$z = \frac{x}{\tan 43°}$$

As $y + z = 80$:

$$\frac{x}{\tan 20°} + \frac{x}{\tan 43°} = 80$$
$$x(\tan 43° + \tan 20°)$$
$$= 80 \times \tan 20° \times \tan 43°$$
$$x = \frac{80 \times \tan 20° \times \tan 43°}{\tan 43° + \tan 20°}$$
$$= 20.943\,251\,98\ldots$$
$$= 21 \text{ (nearest whole)}$$

$\therefore$ the seagull is 21 m high.

4. 50 s

Let the distance be x.

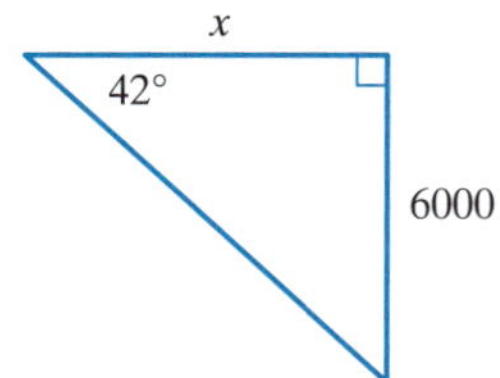

$$\frac{6000}{x} = \tan 42°$$
$$x = \frac{6000}{\tan 42°}$$
$$= 6663.675\,089\ldots$$
$$= 6664 \text{ (nearest whole)}$$

$\therefore$ the plane has flown 6.664 km

$$\text{Time} = \text{Distance} \div \text{Speed}$$
$$= 6.664 \div 480$$
$$= 0.013\,882\,656\ldots \text{ hours}$$
$$= 50 \text{ s (nearest second)}$$

$\therefore$ the plane will take 50 s.

5. 35 km, 110°

The closest point is when $\angle ACB = 90°$.

Let the distance be x.

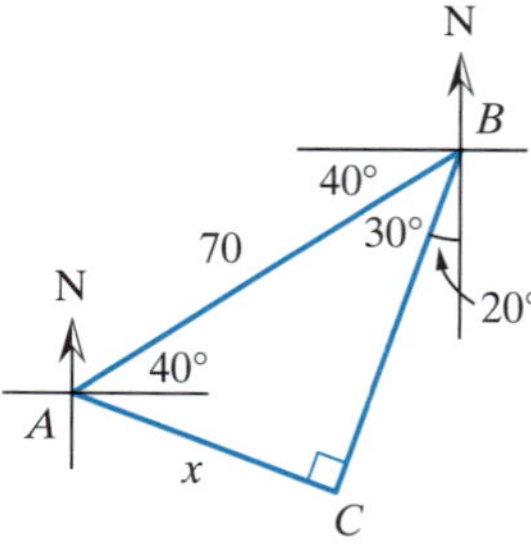

$$\frac{x}{70} = \sin 30°$$
$$x = 70 \times \sin 30°$$
$$= 35$$

$\therefore$ the distance is 35 km.

As $\angle BAC = 60°$, then the bearing of C from A is 110°.

6. 5:44 pm

Let the distances be x and y.

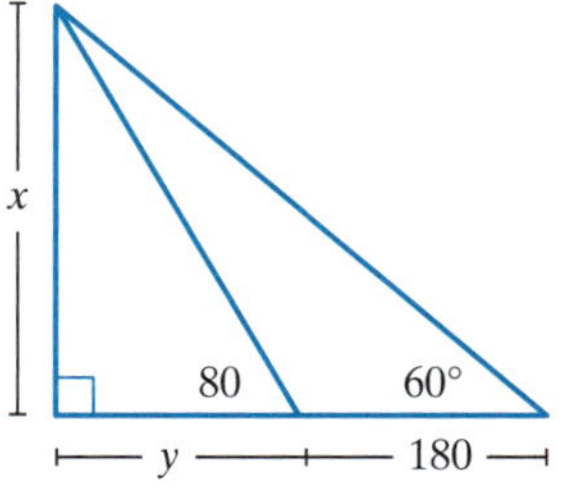

$$\frac{x}{y} = \tan 80°$$
$$x = y \times \tan 80°$$
$$\text{Also, } \frac{x}{y + 180} = \tan 60°$$
$$x = (y + 180) \times \tan 60°$$
$$\therefore y \times \tan 80°$$
$$= (y + 180) \times \tan 60°$$
$$y \times \tan 80°$$
$$= y \times \tan 60° + 180 \times \tan 60°$$
$$y \times \tan 80° - y \times \tan 60°$$
$$= 180 \times \tan 60°$$
$$y(\tan 80° - \tan 60°)$$
$$= 180 \times \tan 60°$$
$$y = \frac{180 \times \tan 60°}{\tan 80° - \tan 60°}$$
$$= 79.144\,671\,74\ldots$$
$$= 79.14 \text{ (2 dec. pl.)}$$
$$\text{Time} = \text{Distance} \div \text{Speed}$$
$$= 79.14 \div 80$$
$$= 0.989\,308\,396\ldots$$
$$= 59 \text{ min (nearest min)}$$

$\therefore$ Myles is south at 5:44 pm.

Key Skill  (pages 118–119)

1. 2, 9

Existing 5 scores: 5, 6, 7, 8, 10

New 7 scores: ?, 5, 6, 7, 8, ?, 10

If lower extreme = 2, then new score of 2:

2, 5, 6, 7, 8, ?, 10

If upper quartile = 9, then new score of 9:

2, 5, 6, 7, 8, 9, 10

$\therefore$ the new scores are 2 and 9.

2. $a = 6$, $b = 8$, $c = 12$, $d = 15$

Lower extreme is 6. $\therefore a = 6$

Median is 11. $\therefore c = 12$

$\therefore$ 6, 8, b, 10, 12, 14, d, 16

Lower quartile is 8. $\therefore b = 8$

Upper quartile is 14.5. $\therefore d = 15$

The scores are 6, 8, 8, 10, 12, 14, 15, 16.

$\therefore a = 6$, $b = 8$, $c = 12$, $d = 15$

3. **15, 18**

Initial 10 scores:

8, 9, 10, 10, 12, 12, 14, 15, 16, 18

New upper extreme is 16, so 18 is ignored:

8, 9, 10, 10, 12, 12, 14, 15, 16

Median drops from 12 to 11, so a score higher than 11 needs to be ignored. If 15 is ignored the upper quartile changes to 13:

$\therefore$ 8, 9, 10, 10, 12, 12, 14, 16

$\therefore$ 15 and 18 were ignored.

4. Median = 60% = 24 out of 40.

As 7 students is 25% of class, then lower quartile is 50% = 20 out of 40.

IQR = 12, then upper quartile is 20 + 12 = 32.

Upper extreme = 95%
= 38 out of 40.

Lower extreme = 38 − 32
= 6

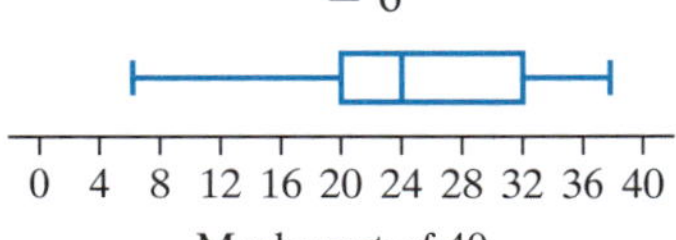

5. **75%**

In Test A 25% of students scored 20 or less, whereas in Test B no-one scored less than 20. In Test A no student scored more than 45, whereas in Test B 50% of students scored 45 or more.

As 25 + 50 = 75, then approximately 75% of students improved their result.

As 75% of 20 = 15, then at least 15 students improved.

6. $\boldsymbol{a = 10, b = 12, c = 14, d = 18, e = 20}$

Lower extreme is 10, $\therefore a = 10$.

Upper extreme is 20, $\therefore e = 20$.

Median is 13, which is the middle of b and c.

Lower quartile is 12, $\therefore b = 12$.

$\therefore c = 14$

Upper quartile is 18, $\therefore d = 18$.

$\therefore a = 10, b = 12, c = 14, d = 18, e = 20$

Key Skill  49 (pages 120–121)

1. $\dfrac{7}{15}$

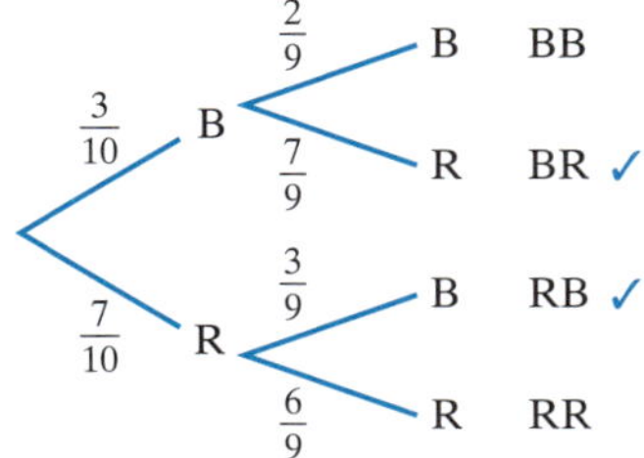

P(different colours)

$= P(BR) + P(RB)$

$= \frac{3}{10} \times \frac{7}{9} + \frac{7}{10} \times \frac{3}{9}$

$= \frac{42}{90}$

$= \frac{7}{15}$

$\therefore$ the probability is $\frac{7}{15}$.

2. $\dfrac{10}{21}$

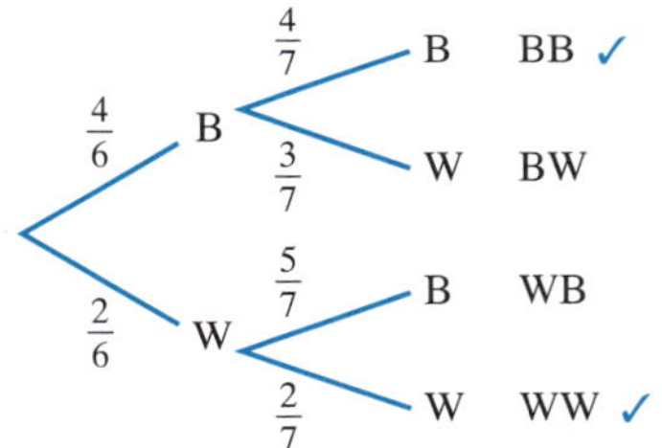

P(same colour) = P(BB) + P(WW)

$= \frac{4}{6} \times \frac{4}{7} + \frac{2}{6} \times \frac{2}{7}$

$= \frac{20}{42}$

$= \frac{10}{21}$

$\therefore$ the probability is $\frac{10}{21}$.

3. $\dfrac{3}{20}$

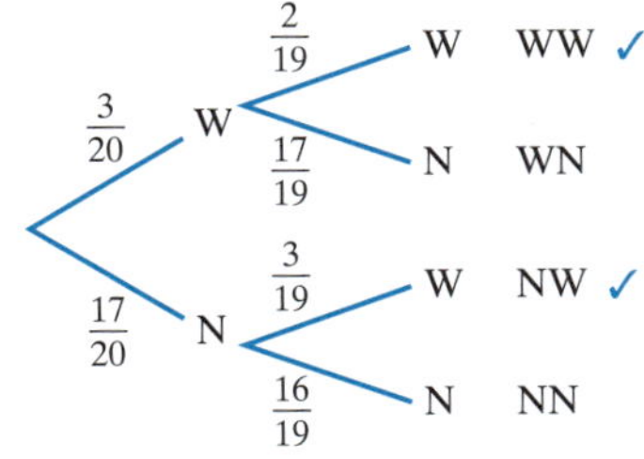

P(wins second prize)

$= P(WW) + P(NW)$

$= \frac{3}{20} \times \frac{2}{19} + \frac{17}{20} \times \frac{3}{19}$

$= \frac{57}{380}$

$= \frac{3}{20}$

$\therefore$ the probability is $\frac{3}{20}$.

4. $\dfrac{95}{126}$

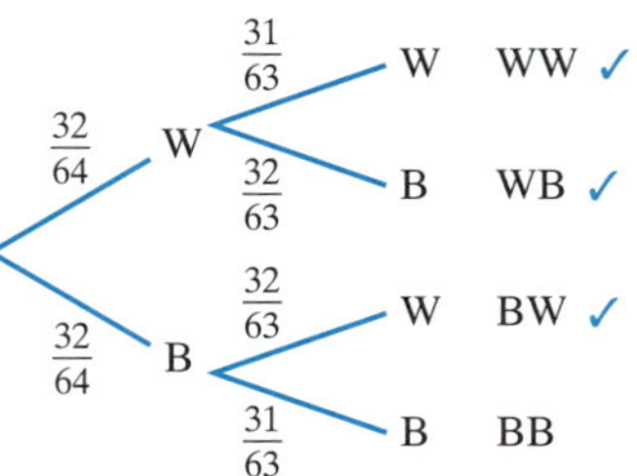

P(at least one white)

= P(WW) + P(WB) + P(BW)

= 1 − P(BB)

$= 1 - \frac{32}{64} \times \frac{31}{63}$

$= \frac{95}{126}$

$\therefore$ the probability is $\frac{95}{126}$.

5. **0.35**

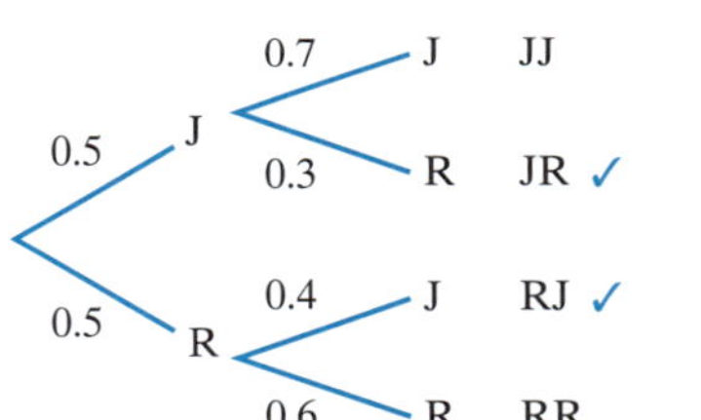

P(Jack wins one game)

= P(JR) + P(RJ)

= 0.5 × 0.3 + 0.5 × 0.4

= 0.35

$\therefore$ the probability is 0.35.

6. **0.03**

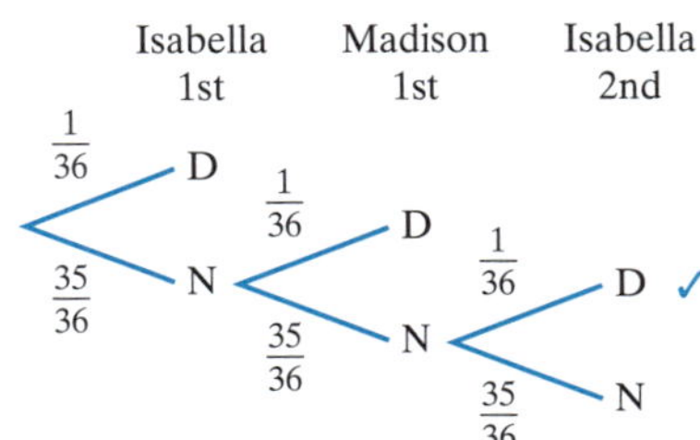

P(Isabella: no double, Madison: no double, Isabella: double)

$= P(NND)$

$= \frac{35}{36} \times \frac{35}{36} \times \frac{1}{36}$

= 0.026 256 001…

= 0.03 (2 dec. pl.)

$\therefore$ the probability is 0.03.

Key Skill 50 (pages 122–123)

1. $\frac{2}{3}$

Sample space = {BB, BG, GB, GG}

P(boy|one is a girl)

$= \text{P(boy \& girl)} \div \text{P(girl)}$

$= \frac{2}{4} \div \frac{3}{4}$

$= \frac{2}{3}$

∴ the probability is $\frac{2}{3}$.

2. $\frac{19}{50}$

		Mode of transport				Total
		Bus	Car	Walk	Bike	
School years	7, 8	24	44	20	12	100
	9, 10	32	38	22	8	100
	11, 12	28	50	16	6	100
Total		84	132	58	26	300

P(car|Year 9 or 10)

$= \text{P(car \& 9 or 10)} \div \text{P(9 or 10)}$

$= \frac{38}{300} \div \frac{100}{300}$

$= \frac{38}{100}$

$= \frac{19}{50}$

∴ the probability is $\frac{19}{50}$.

3. $\frac{10}{29}$

P(Year 7 or 8|walk)

$= \text{P(Year 7 or 8 \& walk)} \div \text{P(walk)}$

$= \frac{20}{300} \div \frac{58}{300}$

$= \frac{20}{58}$

$= \frac{10}{29}$

∴ the probability is $\frac{10}{29}$.

4. $\frac{13}{31}$

Total of 60 students.

P(Chemistry|Biology)

$= \text{P(Chemistry \& Biology)}$

$\div \text{P(Biology)}$

$= \frac{13}{60} \div \frac{31}{60}$

$= \frac{13}{31}$

∴ the probability is $\frac{13}{31}$.

5. $\frac{9}{17}$

Total of 60 students.

P(not Physics|not Chemistry)

$= \text{P(not Physics \& not Chemistry)}$

$\div \text{P(not Chemistry)}$

$= \frac{18}{60} \div \frac{34}{60}$

$= \frac{18}{34}$

$= \frac{9}{17}$

∴ the probability is $\frac{9}{17}$.

6. $\frac{5}{16}$

	Children	Adult	Total
Fiction	250	650	900
Non-fiction	550	550	1100
Total	800	1200	2000

P(fiction|children's book)

$= \text{P(fiction \& children's book)}$

$\div \text{P(children's book)}$

$= \frac{250}{2000} \div \frac{800}{2000}$

$= \frac{250}{800}$

$= \frac{5}{16}$

∴ the probability is $\frac{5}{16}$.

Revision Test  15 (page 124)

1. **2, 4**

Existing 6 scores: 4, 6, 6, 8, 8, 8.
Lower extreme was 4, but from box plot, new lower extreme is 2.
∴ one new score of 2.
Also, median drops from 7 to 6, so the other new score is lower than 6:
Lower quartile drops from 6 to 4, so that the other new score is 4.
The 8 scores: 2, 4, 4, 6, 6, 8, 8, 8.
∴ the new scores are 2 and 4.

2. $\frac{15}{28}$

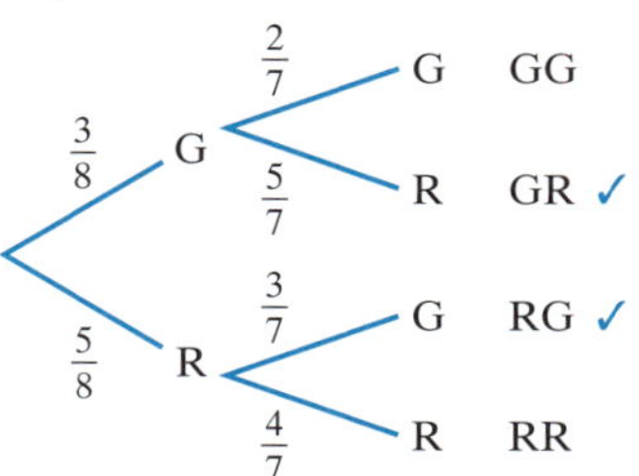

$\text{P(GR)} + \text{P(RG)} = \frac{3}{8} \times \frac{5}{7} + \frac{5}{8} \times \frac{3}{7}$

$= \frac{30}{56}$

$= \frac{15}{28}$

∴ the probability is $\frac{15}{28}$.

3. $\frac{2}{65}$

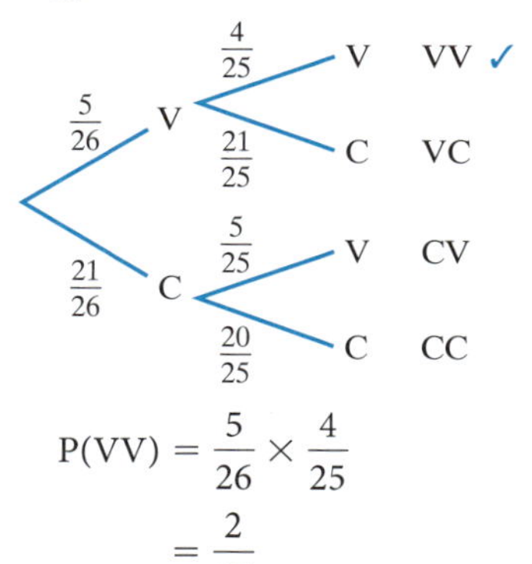

$\text{P(VV)} = \frac{5}{26} \times \frac{4}{25}$

$= \frac{2}{65}$

∴ the probability is $\frac{2}{65}$.

4. $\frac{1}{10}$

Total people surveyed = 40.

P(enjoy cricket but not football)

$= \frac{4}{40}$

$= \frac{1}{10}$

∴ the probability is $\frac{1}{10}$.

5. $\frac{1}{40}$

		Green		
		Yes	No	Total
Red	Yes	28	4	32
	No	7	1	8
	Total	35	5	40

$\text{P(no green, no red)} = \frac{1}{40}$

∴ the probability is $\frac{1}{40}$.

6. $\frac{5}{18}$

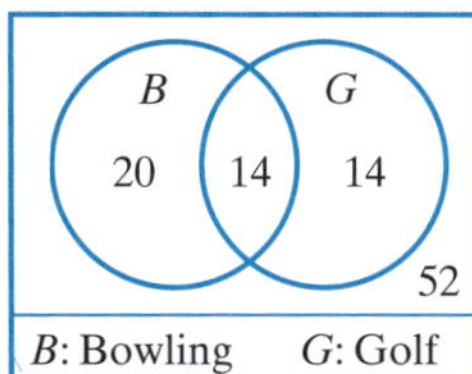

P(bowling|not golf)
$= \text{P(bowling \& not golf)} \div \text{P(not golf)}$
$= \frac{20}{100} \div \frac{72}{100}$
$= \frac{20}{72}$
$= \frac{5}{18}$

$\therefore$ the probability is $\frac{5}{18}$.

Revision Test 16 (page 125)

1. $p = 7, q = 9, r = 11, s = 13, t = 15$
Lower extreme is 7, $\therefore p = 7$.
Upper extreme is 15, $\therefore t = 15$.
Lower quartile is 9, $\therefore q = 9$.
Upper quartile is 13, $\therefore s = 13$.
Median is 12, which is the middle of r and 13.
$\therefore r = 11$.
$\therefore p = 7, q = 9, r = 11, s = 13, t = 15$.

2. $\frac{23}{50}$

	Glasses	No glasses	Total
Boys	8	12	20
Girls	7	23	30
Total	15	35	50

P(girl, no glasses) $= \frac{23}{50}$

$\therefore$ the probability is $\frac{23}{50}$.

3. 0.36
P(TT) = 0.16
P(T) $= \sqrt{0.16}$
$= 0.4$
P(H) $= 1 - 0.4$
$= 0.6$
P(HH) $= 0.6^2$
$= 0.36$
$\therefore$ the probability is 0.36.

4. $\frac{4}{9}$

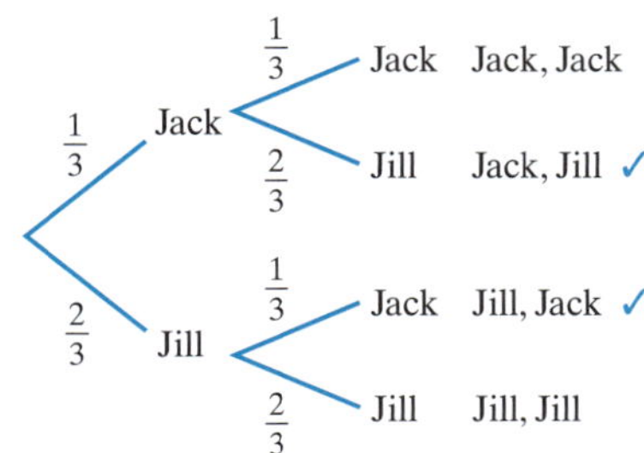

P(one game each)
= P(Jack, Jill) + P(Jill, Jack)
$= \frac{1}{3} \times \frac{2}{3} + \frac{2}{3} \times \frac{1}{3}$
$= \frac{4}{9}$

$\therefore$ the probability is $\frac{4}{9}$.

5. $\frac{22}{51}$
8 green, 4 red, 6 blue.
P(not blue) $= \frac{12}{18}$
P(not blue, not blue) $= \frac{12}{18} \times \frac{11}{17}$
$= \frac{22}{51}$

$\therefore$ the probability is $\frac{22}{51}$.

6. $\frac{19}{68}$

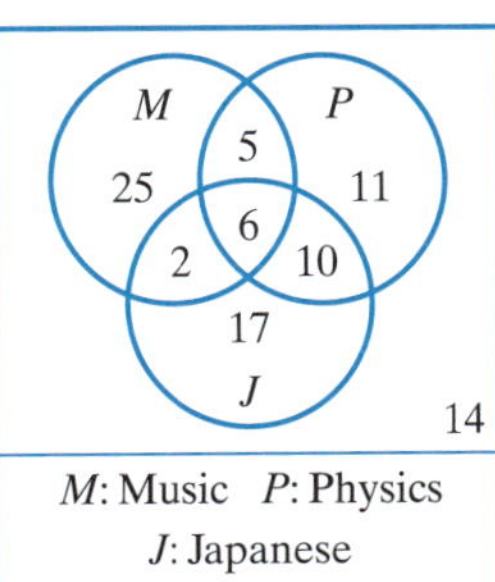

P(Japanese|not Physics)
$= \text{P(Japanese \& not Physics)} \div \text{P(not Physics)}$
$= \frac{19}{100} \div \frac{68}{100}$
$= \frac{19}{68}$

$\therefore$ the probability is $\frac{19}{68}$.

NOTES